Properties of Liquids and Solutions

Properties of Liquids and Solutions

Second Edition

J. N. Murrell and A. D. Jenkins
University of Sussex, Brighton, UK

JOHN WILEY & SONS

Chichester · New York · Brisbane · Toronto · Singapore

Copyright © 1982, 1994 by John Wiley & Sons Ltd,
Baffins Lane, Chichester,
West Sussex PO19 1UD, England

Telephone National Chichester (0243) 779777
International +44 243 779777

Other Wiley Editorial Offices

John Wiley & Sons, Inc., 605 Third Avenue,
New York, NY 10158-0012, USA

Jacaranda Wiley Ltd, 33 Park Road, Milton,
Queensland 4064, Australia

John Wiley & Sons (Canada) Ltd, 22 Worcester Road,
Rexdale, Ontario M9W 1L1, Canada

John Wiley & Sons (SEA) Pte Ltd, 37 Jalan Pemimpin #05-04,
Block B, Union Industrial Building, Singapore 2057

Library of Congress Cataloging-in-Publication Data

Murrell, J. N. (John Norman)
    Properties of liquids and solutions / J. N. Murrell and
A. D. Jenkins. — 2 nd ed.
        p.   cm.
    Includes bibliographical references and index.
    ISBM 0-471.-94418-1 (cloth) — ISBM 0-471-94419-X (pbk.)
    1. Liquids.   2. Solution (Chemistry)   I. Jenkins, A. D. (Aubrey
Dennis)   II. Title.
    QD541.M9   1994
    540′.0422—dc20   93-46721
                                                                              CIP

British Library Cataloguing in Publication Data

A catalogue record for this book is available from the British Library

ISBN 0 471 94418 1 (cloth)
ISBN 0 471 94419 X (paper)

Typeset by Keytec Typesetting Ltd, Bridport, Dorset
Printed in Great Britain by Biddles Ltd, Guildford, Surrey

Contents

Preface to the First Edition

This book has been based upon a course of lectures given to second year students in the School of Chemistry and Molecular Sciences at the University of Sussex. To understand our choice of material and the level at which it is presented we should explain that certain courses at Sussex have been identified as core material which is given to all students in the School: chemists, biochemists, chemical physicists, materials scientists and environmental scientists.

The reasons for choosing *Properties of Liquids and Solutions* as a core course, are, we hope, obvious. Most chemical synthesis is carried out in the liquid state because reactions are on the one hand faster than in the solid state and on the other hand energetically cheaper and mechanistically simpler than in the gaseous state. Most methods of purification involve either a distribution of solutes between immiscible solvents or the partition of mixtures in solid–liquid or liquid–gas equilibria. Most biochemical and many environmental systems are in the liquid state or involve colloids or membrane boundaries between liquids. Solid–liquid equilibria are particularly important for the industrial preparation and purification of materials.

We have written the book because we found no single text that met the needs of our students. Most of the material we present can be found either in general physical chemistry texts or in monographs. However, the former generally lack the detail that is required for a course at honours degree level and the latter lack the cohesion that comes from teaching the subject as a whole and are often too advanced in their speciality. There is no shortage of suitable texts to support undergraduate teaching in, for example, spectroscopy, reaction kinetics, thermodynamics or valence theory. The absence of a comprehensive text on the liquid state probably means that it is not generally taught as a unified subject. We hope that this book will encourage others to look at it in this light.

Much of our knowledge of the liquid state derives from the physical chemistry of the nineteenth century. This is firmly based on thermodynamics and is no less important now than it was fifty years ago. However, with the development of fast, large-store digital computers, our understanding of liquid properties in terms of intermolecular potentials has leapt forward making this currently one of the most active areas of research. We have tried to achieve a balance in this book between the traditional or macroscopic approach and the more modern, and complementary, microscopic approach to liquid properties.

In order to produce a text of moderate length we have assumed a prior knowledge in several important fields. First thermodynamics: we assume that all the basic thermodynamic equations have been covered elsewhere and we expect students to have had some practice in applying them, say to ideal gases or solids.

From statistical mechanics we assume prior knowledge only of the Boltzmann distribution and the partition function. There is a difficult choice here; statistical mechanics has led to substantial advances in our understanding of liquids but these are either associated with simple models of liquids that are now thought inadequate or they are connected with integral equations which we would judge to be largely outside the undergraduate curriculum. We do describe in detail the statistical approach through computer simulation because it is modern, not mathematically demanding and conceptually revealing; all very positive attributes of undergraduate material.

Finally, from quantum mechanics we require a few elementary concepts of bonding theory such as orbital, spin and covalent, which will enable us to give a review of intermolecular forces and provide the foundation of the microscopic approach.

The order in which we have chosen to cover the subject is broadly based on a pattern of increasing complexity; non-electrolytes before electrolytes before colloids, for example. We have also chosen to cover the theoretical material on intermolecular forces and liquid state models before the more experimental aspects of pure fluids and non-electrolyte mixtures. However, the reader would lose nothing by taking Chapters 4 and 6 before Chapters 2 and 3 and we would expect many courses to follow that pattern.

We are indebted to many of our colleagues for stimulating discussions during the preparation of the book, particularly Drs M. H. Ford-Smith, J. B. Pedley, G. L. Pratt and J. G. Stamper. We are particularly grateful to Drs E. R. A. Peeling and M. J. B. Evans for having read and criticized parts of the manuscript.

Preface to the Second Edition

The first edition of this book was written more than ten years ago and in the history of the subject (Clapeyron's equation was discovered in 1834) this is a very short time. However, our views on several branches of the subject are still changing rapidly as we acquire a better understanding of intermolecular potentials and make more extensive computer simulations of bulk properties.

The first edition has been out of print for three years, and in response to enthusiastic comments from readers and reviewers we have updated several parts of the text and extended a few topics (e.g. introduced a short examination of supercritical fluids). Chapters 11 and 13 have been rewritten; to reflect a new perspective on the subject rather that to present new material.

We would like to acknowledge the substantial contributions of Dr Ernest Boucher to the first edition of the book. He was an eminent scholar in several of the traditional branches of the subject; capillarity, wetting, adsorption, etc., and we regret that his death in 1988 severed his valuable link with the book.

Chapter 1

The Liquid State of Matter

1.1 WHAT IS A LIQUID?

A simple operational definition of a liquid is that it is a medium which takes the shape of a container without necessarily filling it. In contrast, a gas both takes the shape and fills the container whilst a solid neither takes the shape nor fills it. There are, however, materials which are difficult to place on this definition. A few polymeric materials appear to be solid but with time they will flow to the shape of a container. Glasses behave similarly, although their flow rate may be immeasurably slow at normal temperatures.

The operational distinction between a solid and a liquid must therefore be made more precise by referring to the time-response to an applied force. We say that a material shows elastic behaviour if it returns to its original shape following the application then removal of a force. It shows inelastic or plastic behaviour if it is permanently distorted by the force.

The response of solids to a force is mainly elastic (at least for low strain), and that of liquids is mainly inelastic. However, the skipping of stones on the surface of water shows that elasticity is not completely absent in a liquid and the flowing of glass is an illustration of inelastic behaviour. The polymer known colloquially as 'silly putty', which can be moulded into an elastic ball but which will flow slowly under an applied force, is a material in which both elastic and inelastic behaviour are easily illustrated.

At first sight the operational distinction between a liquid and a gas might appear to be more rigorous. However, there is a simple experiment to show that this is not so.

If one takes a sealed tube containing liquid in equlibrium with its vapour, an interface between the phases is clearly visible (the meniscus). If the temperature of the tube is raised, the density of the liquid will decrease and that of the gas increase until eventually a temperature is reached at which both densities will be equal. At this point all distinction between liquid and vapour disappears and this is shown by the disappearance of the interface.†

The temperature at which this occurs is called the critical temperature. As we shall see later the pressure in the tube is also a characteristic property of the liquid which is called the critical pressure. For most liquids this pressure is very high (e.g. for water it is 218 atm) so that great care is needed in carrying out the experiment.

On applying increasing pressure to a substance at a temperature above its critical temperature it simply becomes more dense: there is no point at which one can say it passes from being a gas to being a liquid. This unification of the two states of matter is expressed by using the term 'fluid' to encompass both gas and liquid. Below the critical point gas and liquid are distinct states which can exist in equilibrium with each other. Above the critical point the two states become one.

Supercritical fluids are being increasingly used for extraction processes. For example, supercritical CO_2 is being used commercially to decaffeinate green coffee beans. The advantages of such a process are that CO_2 is pharmacologically inert, and any extract can be recovered simply by lowering the pressure. Supercritical fluids will be examined in more detail in Chapter 6.

A further brief comment on terminology: the words vapour and gas are often used interchangeably. Vapour is usually used when the phase is in contact with its liquid or solid. For example vapour pressure is the equilibrium pressure in the gas (vapour) above a liquid or solid. One could logically always use vapour for the phase below the critical temperature because only then can this equilibrium be established. However, we shall generally use the term gas unless it is essential to make the distinction that the substance can be placed in equilibrium with its liquid by a change in pressure alone.

Macroscopic and Microscopic Views

Although there is no simple and all-embracing operational distinction between a solid, liquid and gas, ice, liquid water and water vapour are obviously different and hence more scientific lines of investigation must be pursued to penetrate their differences. We shall see that progress can be made both by thermodynamic and molecular descriptions of materials, approaches which can be called the macroscopic and microscopic views, respectively. Historically the thermodynamic approach has precedence as it

---

†In 1863 Andrews described the phenomenon for carbon dioxide in these words: 'On partially liquefying carbonic acid by pressure alone and gradually raising at the same time the temperature to 88 °F the surface of demarcation between the liquid and the gas became fainter, lost its curvature and at last disappeared. The space was then occupied by a homogeneous fluid, which exhibited, when the pressure was suddenly diminished or the temperature slightly lowered, a peculiar apearance of moving or flickering striae throughout its entire mass.' The latter observation which is associated with opalescence of the fluid is due to fluctuations in the density of the fluid near the critical point.

stems from the work of Willard Gibbs and others in the late nineteenth century. The molecular approach to liquids is largely based on theories developed in the past fifty years, but the major advances in our understanding of liquids that it has brought are directly linked to the advent of the electronic computer in the 1950s.

The strength of classical thermodynamics lies in the fact that its application does not rely on any knowledge of the molecular structure of matter. In fact some important generalizations about the freezing and boiling of liquids were discovered before the atomic structure of matter was universally accepted. However, the diversity of liquid-state properties can only be understood from the diversity of the shapes of molecules and the forces that act between molecules. In other words, there is probably more to be gained from the microscopic approach if we wish to understand the properties of individual systems, but the macroscopic approach leads more easily to the broad generalizations regarding the states of matter. Both approaches have as their aim the prediction and interpretation of measurable quantities.

1.2 AN INTRODUCTION TO PHASE DIAGRAMS

A convenient method of representing the regions of stability of solid, liquid and gas under various conditions of temperature and pressure is by a phase diagram. The formal definition of the term phase was first given by Gibbs. It is a state of matter that is uniform in both its physical state and its chemical composition. Liquid water is a single phase, a mixture of ice and water has two phases. A gas, whether composed of a single chemical substance or a mixture of chemicals, is a single phase. Two different chemicals in the liquid state may form a single phase (the component liquids are said to be miscible) or form two phases; hexane and water, for example, are immiscible.

At one atmosphere external pressure water exists as a solid below 273 K (the ice point), as a liquid between 273 K and 373 K, and as a gas above 373 K. If the pressure is increased then the melting point of ice is lowered and the boiling point of water is raised. If the pressure is reduced below one atmosphere these trends are reversed and below 610 Pa (0.06 atm) ice passes directly to water vapour when the temperature is raised, without going through the liquid phase.† The term sublimation describes the solid–vapour conversion in either direction.

---

†The diversity of units currently in use for pressure causes difficulty for readers of the scientific literature. The SI unit is the pascal which is the pressure exerted by a force of one newton per square metre; $1 Pa = 1 N m^{-2}$. Unfortunately this unit is not as commonly used as many other . SI units. An important and widely used unit is the atmosphere (atm) which is defined by international agreement to be exactly 101 325 Pa. A useful rule of thumb is $10^5 Pa \sim 1$ atm, and since 1982 10^5 Pa has been the internationally accepted standard-state pressure for

4

These data on water are summarized in its phase diagram, Figure 1.1. For simplification we have shown an incomplete diagram which does not record the fact that there are several different crystalline forms of ice. A further examination of the solid phases will be made in Chapter 8.

Figure 1.1 shows *areas* which are two-dimensional, *lines* which are one-dimensional and a *point* (at the joining of the lines) which has zero dimensions. The areas indicate the ranges of temperature and pressure within which only a single phase exists; within the boundaries indicated we

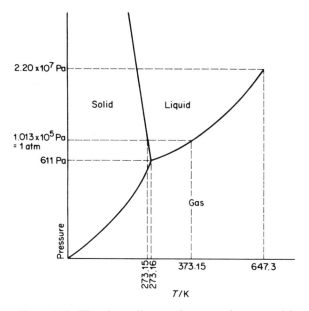

Figure 1.1. The phase diagram for water (not to scale)

thermodynamic purposes. Precisely 10^5 Pa also defines another unit, the bar (hence 1 bar ~ 1 atm) and high pressures are often quoted in kilobar (kbar). The atmosphere in turn is also standardized to the pressure of a column of mercury 760 mm high (1 atm ≡ 760 mmHg). Another name for a standard mmHg is torr, after Torricelli who invented the barometer (1 torr = (1/760) standard atmosphere). In this book the pascal will be generally used for all quantitative measures of pressure but the phrase 'at one atmosphere' will be used to indicate the common laboratory conditions of a system under variable natural atmospheric pressure.

It is quite common to write the ideal gas equation $pV = nRT$ using a lower case letter for pressure but upper case for temperature and volume. The reasoning is that volume is an extensive quantity (its value for the whole is the sum of values for the separate parts) and along with other extensive quantities, like energy and entropy, it is given a capital letter. Pressure, in contrast, is an intensive quantity (its value does not depend on the size of the system) and a lower case letter is used. Temperature is also an intensive quantity but the logical use of a lower case symbol could cause confusion with the use of t for time. We prefer to use capitals for pressure, volume and temperature and we will therefore write the ideal gas equation $PV = nRT$. We shall, however, conform to the common practice of using a lower case letter for vapour pressure when we wish to distinguish this from the total pressure. The vapour pressure is approximately equal to the partial pressure of a component in the vapour above the condensed phase of that component. We shall therefore use a lower case for partial pressures. Thus, Dalton's law of partial pressures will be written $P = \sum_i p_i$.

can vary temperature and pressure whilst maintaining a single phase. We say that with a single phase the system has two degrees of freedom. The number of degrees of freedom is formally defined as the number of variables that can be changed independently without changing the number of phases.

Two phases can coexist in equilibrium only at pressures and temperatures defined by the lines shown in Figure 1.1: the solid–liquid line, the liquid–gas line and the solid–gas line. For two phases in equilibrium, if we specify one of the variables, say pressure, then the other, temperature, is determined. Along a line pressure is a function of temperature or, vice versa, temperature is a function of pressure. Along a line the system has only one degree of freedom.

The vapour pressure of a condensed phase is the pressure at which that phase is in equilibrium with its vapour. Thus the liquid–gas line gives the vapour pressure of the liquid as a function of temperature. If the external pressure is taken below the vapour pressure the liquid will boil and if it is above the vapour pressure the vapour will condense.

Three phases can coexist in equilibrium only at the intersection of the three two-phase lines, which is called a *triple point*. This is a fixed or invariant point which in the phase diagram of water has coordinates $P = 610\,\text{Pa}$, $T = 273.1600\,\text{K}$. Because the position of the triple point is capable of being determined with great accuracy (much greater than that of the ice point, the melting point of ice at 1 atm pressure, because that depends on maintaining the pressure at exactly 1 atm) it has been taken as the fixed point in the Kelvin scale of temperature.

The Phase Rule

The number of degrees of freedom f available to a single component such as water can be summarized mathematically by the formula

$$f = 3 - p \tag{1.1}$$

where the system has p phases in equilibrium. This is a simplified expression of the Gibbs phase rule which is valid for a pure substance. The complete rule, for many components, will be discussed in Chapter 7.

Common experience might be thought to conflict with the pattern of Figure 1.1. On a winter's day we may see ice floating in a pond and sense that the air is saturated with water vapour. Can this mean that the conditions represent those of the triple point? The phase diagram that we have been examining applies only to a single component in an equilibrium situation and the winter's scene is neither. If we assumed equilibrium and ignored the very small dependence of vapour pressure on total pressure then the air pressure could be discounted and the partial pressure of water vapour above an icy pond would equal the triple point pressure of 610 Pa.

The fact that at a specified pressure pure ice has a definite melting point and pure water a definite boiling point is a consequence of the phase rule;

there is only one degree of freedom for a single-component two-phase system. The importance of such results provides the incentive to formalize the phase rule by a proof which we give here for a single component.

Two or more phases can exist in equilibrium only if certain criteria are satisfied. For thermal equilibrium each phase must be at the same temperature; if this were not the case, heat would flow from the hotter to the colder until thermal equilibrium was reached. For mechanical equilibrium each phase must be at the same pressure; a system which does not satisfy this criterion will undergo volume changes until the pressures are equal. Finally, there is the criterion arising from the second law of thermodynamics that the entropy of the system must be a maximum; for a closed system at constant temperature and pressure, this is equivalent to the statement that the Gibbs free energy of the system must be a minimum. The free energy of one mole of a substance G_m (also known as its chemical potential μ, see pages 78, 113) must obviously have the same value in all phases in equilibrium, otherwise material would convert from the phase with higher G_m to the phase with lower G_m, (see also Chapter 4). Thus, for the coexistence of two phases α and β, we may write

$$G_m(\alpha)(T, P) = G_m(\beta)(T, P) \tag{1.2}$$

reminding ourselves that the molar free energy is a function of both temperature and pressure. Figure 4.1 shows that, at a given pressure, there is only a single temperature at which the molar free energies of two phases coexisting in equilibrium have the same value, and there is therefore a restriction on the conditions attaching to coexistence that amounts to the removal of one of the degrees of freedom possessed by a single phase: if the temperature is fixed, the pressure must automatically be fixed, and *vice versa*.

If it is specified that three phases, α, β and γ, are to coexist at equilibrium, there is an additional restrictive condition to be met, i.e. that

$$G_m(\alpha)(T, P) = G_m(\gamma)(T, P) \tag{1.3}$$

which implies that

$$G_m(\beta)(T, P) = G_m(\gamma)(T, P) \tag{1.4}$$

and there is a loss of one more degree of freedom compared with a single phase: thus we arrive at the conclusion that there are no degrees of freedom at all at a triple point.

This result can be generalized by stating that, for a one-component system of p phases in equilibrium, there will be $(p - 1)$ restrictive equations, such as (1.2), (1.3) and (1.4), and hence (p − 1) less degrees of freedom than for a single phase. As a single phase has two degrees of freedom (T and P), the number of degrees of freedom f available to p phases must be

$$f = 2 - (p - 1) = 3 - p \tag{1.5}$$

which is a proof of (1.1).

The Critical Point

The high-temperature end of the liquid/gas line in a phase diagram terminates at the critical point. For water (Figure 1.1) the parameters of the critical point are $T_c = 647.30\,\text{K}$ and $P_c = 2.20 \times 10^7\,\text{Pa}$. Table 1.1 shows a selection of critical-point data. Note that critical temperatures occur in a very wide range from a few degrees kelvin for helium to above $10^4\,\text{K}$ for some metals. The critical temperature is roughly a constant multiple of the normal boiling point, $T_c \approx 1.5 T_b$, and we will comment on this fact in a later chapter see pp 63, 67). Critical pressures also cover a wide range, and systems which are liquid at room temperature and atmospheric pressure normally have critical pressures above 100 atm.

Figure 1.1 brings out the points we have made about the unification of the liquid and gaseous phases above the critical point. It is also clear that by taking a suitable (P, T)-path in the phase diagram it is possible to pass from a point in which the phase is gaseous to one in which it is liquid without crossing the gas–liquid line. All that is necessary is, first, to raise the temperature keeping the pressure constant until T_c is exceeded, second, to increase the pressure keeping the temperature constant until P_c is exceeded and, finally, to reduce the temperature below T_c.

Density and Compressibility

A liquid under normal pressures has a density which is not too dissimilar from that of the solid; this is true over the whole liquid range from the melting point to the boiling point. Generally the liquid is less dense than the solid at the melting point, but there are a few exceptions of which water is one; ice floats on water. Silicon, germanium and tin are other examples where the liquid is more dense than the solid. This can be explained by the fact that the solids have rather open crystal structures with the low coordination number of four.

Another similarity between normal liquids and solids is that they have a low compressibility, which is an important property for the successful operation of hydraulic transmission systems. We can infer from this that

Table 1.1. Values of some critical constants for fluids

	$P_c/10^5\,\text{Pa}$	T_c/K	$V_{c,m}/\text{cm}^3\,\text{mol}^{-1}$
He	2.29	5.2	57.8
H_2	13.0	33.2	65.0
N_2	33.9	126	90.1
CO_2	73.9	304	94.0
C_2H_6	48.8	305	148
C_6H_6	49.2	562	260

Data from *handbook of Chemistry and Physics*, 59th Edn, CRC Press (1978).

there is not a great deal of free space between the molecules in a normal liquid and in this respect they resemble the solid phase.

The above statements about density and compressibility do not apply to liquids near their critical points which are much less dense and quite compressible. Water, for example, has a density of only $0.32\,\mathrm{g\,cm^{-3}}$ at its critical point so that it expands in volume by about a factor of three on going from its normal melting point to the critical point. A liquid closer to its critical point has more free space between the molecules than does the normal liquid.

Crystalline Solids

There is no evidence that any substance has a critical point on the solid–liquid boundary line. No matter how high the pressure there is always a sharp distinction between liquid and solid and a discontinuous change in volume on passing across the phase boundary. This result does not conflict with our earlier statement that it is difficult to categorize glasses and some plastic materials as solids or liquids. A phase diagram represents the regions of existence of thermodynamically stable species. Systems can be kinetically stable in the sense that they do not change during the time-scale of an experiment, but they are not then subject to the laws of equilibrium systems. It is, for example, possible to cool a liquid below its freezing point without it freezing (called supercooling) providing there is an absence of nucleating centres. Liquids that become very viscous close to their freezing point are very slow to crystallize and will, on further lowering of the temperature, become glasses.

The thermodynamic distinction we have to make, therefore, is not between a liquid and a solid, but between a liquid and a *crystalline* solid. A crystal is defined as a material which possesses a periodic repetition of some basic microstructure, this repetition extending three-dimensionally over macro-scopic distances. We say that a crystal is a macroscopically ordered system, and specifically the molecules in a crystal possess long-range order in both the positions of their centres and their orientations. Gases and liquids do not possess this long-range order but, as we shall see in the next section, liquids are not devoid of all order.

1.3 ORDER IN LIQUIDS

Dilute gases are without order in the spatial arrangement of their molecules. If the probability per unit volume of finding a molecule in the gas with its centre at the point r in space is $P(r)$, then the probability per unit volume of finding simultaneously one molecule at r and another at s will be

$$P(r, s) = P(r) \times P(s) = P^2(r) \tag{1.6}$$

providing that the distance between the two molecules, $|r - s|$, is larger than the sum of their van der Waals radii. In other words, providing that the two molecules are not attempting to occupy the same region of space, their probabilities for being at particular points are uncorrelated.

Figure 1.2 shows schematically the pair probability function for a dilute gas and we can take this as being typical of a phase with no order. This may be contrasted with the probability function for a close-packed crystal lattice of spherical molecules which is shown in Figure 1.3. For this the pair probability function shows peaks and for an ideal lattice these peaks persist to infinite distance.

Diffraction

It is well known that a crystal lattice will produce a diffraction pattern from scattered monochromatic radiation with a wavelength of a similar order of magnitude to the lattice spacing in the crystal. By analysis of this pattern the

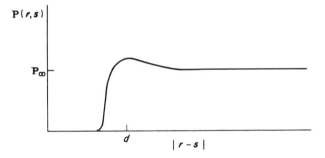

Figure 1.2. The pair probability function for a dilute gas as a function of the distance between their centres. The van der Waals diameter of the gas molecule is d, and P_∞ is a constant that depends on the density of the gas. The maximum at d is due to the attractive forces that exist between molecules (see Chapter 2). The higher the temperature, the less prominent will be this maximum

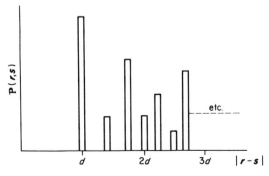

Figure 1.3. The pair probability function for a cubic close-packed solid. The first peak corresponds to 12 nearest neighbours at a distance d

distances between the atoms in the crystal can be determined. Both X-rays and mono-energetic neutrons (with a wavelength λ given by the de Broglie relationship $\lambda = h/p$ where h is Planck's constant and p is the momentum) can be obtained with wavelengths of the order of 0.1 nm (1 Å) and these are suitable for such diffraction studies.

A monatomic gas (e.g. Ne) gives rise to no diffraction pattern, showing that there is no regular spacing between the atoms. Molecular gases do show diffraction,† but this is due to the regular spacing between the atoms within a molecule and not between atoms in different molecules. Liquids occupy an intermediate position between the crystalline solid and the gas. They do show diffraction as a result of regular spacing between atoms in different molecules but the diffraction peaks are less sharp than those procduced by a solid.

A rigorous derivation of the diffraction formula for a liquid goes beyond the mathematical level of this book but an outline of it can be given. Consider first a monochromatic beam scattered by two centres A and B through an angle θ as shown in Figure 1.4. Beam (1) will travel further than beam (2) by $\delta = r \sin \theta$, before it reaches the detector. If this distance is an integer number of wavelengths ($\delta = n\lambda$) the two beams will arrive at the detector in phase and an amplified signal will be detected. If the path distances differ by an extra half wavelength ($\delta = (n + \frac{1}{2})\lambda$) the two beams will be out of phase and there will be no signal. The detector will then show a series of maxima and minima as one sweeps through the angle θ, analogous to the diffraction pattern produced by visible light from a double slit.

Suppose that AB is a diatomic molecule; δ will then also depend on the angle (α) between AB and the incident beam. It might be thought that averaging over all angles α would eliminate the diffraction pattern but this is

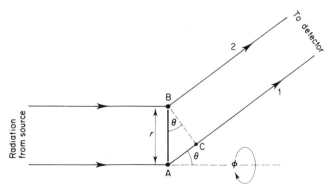

Figure 1.4. Scattering of a beam by two scattering centres A and B where AB is perpendicular to the beam

---

†This is most readily seen in electron scattering rather than in X-ray or neutron scattering as the scattering probability is much greater for electrons.

not so because there is more chance of finding molecules with large α than with small. If all orientations in space are equally probable then the number of molecules making an angle α with the incident beam is proportional to $\sin \alpha$. It can be shown that after averaging over α the intensity of the scattered beam is given by the expression

$$l(s) \propto \frac{\sin (sr)}{sr}, \tag{1.7}$$

where

$$s = \frac{4\pi}{\lambda} \sin \frac{\theta}{2}. \tag{1.8}$$

Sin (sr) is an oscillating function hence $I(s)$ will oscillate and decay with increasing s or θ. All azimuthal angles ϕ relative to the incident beam are equally probable so that the diffraction pattern will consist of concentric rings around a centre which is the undeflected beam, these rings gradually dying out for large angles. This would be the typical electron diffraction pattern from a gas of diatomic molecules.

The Radial Distribution Function

If we turn now to a monatomic liquid then we no longer have a fixed distance r between the scattering centres.† The spread of possible distances can be described by a probability function which defines mathematically the chance of finding two atoms separated by r. This function is called the radial distribution function $g(r)$. To be more precise it is defined such that from one atom there exists a second atom at a distance between r and $r + dr$ with probability $4\pi r^2 g(r)\, dr$.

The intensity of the scattered X-ray or neutron beam produced by a liquid is obtained by multiplying equation 1.7 by the above probability function and integrating over all values of r, thus

$$I(s) = A \int_0^\infty 4\pi r^2\, g(r)\left(\frac{\sin sr}{sr}\right) dr, \tag{1.9}$$

where A is an appropriate constant which depends on the scattering strength of the atoms and beams in question. Expression 1.9 enables us to calculate the scattering pattern knowing the radial distribution function. In practice there is more interest in determining $g(r)$ from the experimentally determined scattering pattern.

Expression 1.9 can be rewritten as the following Fourier integral

$$sI(s) = A \int_0^\infty [4\pi r\, g(r)] \sin (sr)\, dr, \tag{1.10}$$

---

†Even for a diatomic molecule r is not absolutely fixed due to vibrational motion; the short-time average of r is, however, fixed.

12

and using the inverse Fourier theorem this can be transformed into

$$4\pi r\, g(r) = \frac{2}{A\pi} \int_0^\infty sI(s) \sin(sr)\,ds, \tag{1.11}$$

whence

$$g(r) = \frac{1}{2A\pi^2 r} \int_0^\infty sI(s) \sin(sr)\,ds. \tag{1.12}$$

Thus knowing $I(s)$ and the constant A it is possible to determine $g(r)$.

One of the most striking illustrations of the difference between a solid and a liquid is a comparison of the X-ray diffraction of the two phases close to the melting point. Figure 1.5 shows this for powdered and liquid aluminium near the melting point. Analysis of the diffraction patterns leads to the radial distribution functions also shown in Figure 1.5. Notice that the sharp pattern of the powder gives on analysis a distribution function which also has

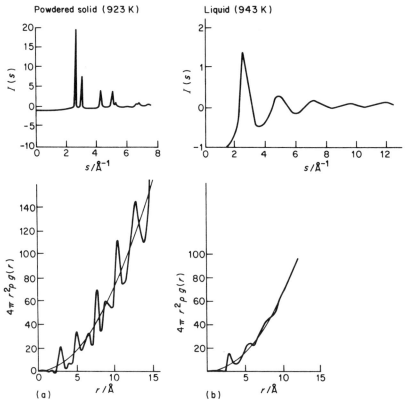

Figure 1.5. The X-ray diffraction intensities from powdered and liquid aluminium close to the melting point. $4\pi r^2 \rho g(r)$ is the probability of finding two atoms separated by a distance r in a material of density ρ (From H. Ruppersberg and H. J. Seeman, Z. *Naturforsch.* **20a**, 104 (1965))

relatively sharp oscillations and these persist to large distances. The liquid, in contrast, has a smoother diffraction pattern and only the first two oscillations in $g(r)$ are prominent. On the evidence, there is order in liquid aluminium as far as second neighbours but not significantly beyond that.

We will return to the question of order in a liquid later in the book with particular reference to models of the liquid state, and to liquid crystals which occupy a position between the liquid and the crystalline states.

1.4 CLASSES OF LIQUIDS

In this book we emphasize the principles that guide our understanding of all liquids. However, one must not understate the difference between one liquid and another. Helium, aluminium, water and benzene in their liquid states have widely different properties; the corresponding solids are also very different from one another. These differences can be attributed to differences in the intermolecular potentials in the condensed phases, and we shall therefore preface our discussion of different classes of liquids and liquid mixtures by reviewing, in the next chapter, the theory of interatomic and intermolecular forces.

Table 1.2 lists the phase transition temperatures for the four liquids just mentioned, and two other properties, density and viscosity. Liquid helium has unique properties† because the lightness of the atoms and the weakness of the He–He interatomic potential lead to a fluid with properties which must be intrrerpreted by quantum mechanics; for all other fluids classical mechanics is generally sufficient.‡

Figure 1.6 shows the phase diagram for helium at low temperatures. There are in fact two isotopes of helium, ^{3}He and ^{4}He, which have been studied separately and we show only the phase diagram of the common

Table 1.2. Properties of some widely different liquids

	He	C_6H_6	H_2O	Al
T_m/K	0 (25 atm)	279	273	933
T_b/K	4.21	353	373	2740
T_c/K	5.20	562	647	7740
$\rho(T_m)/g\ cm^{-3}$	0.125 (4.2 K)	0.899	1.000	2.380
$\eta(T_m)/cP$	0.003 (4.2 K)	0.85	1.79	1.39

T_m, T_b and T_c are the melting, boiling and critical temperatures, respectively. ρ is the density and η the viscosity at the melting point (for He the boiling point). A convenient unit of viscosity is the centipoise, $cP = 10^{-3}\ kg\,m^{-1}\,s^{-1}$.

---

†J. Wilks. *An Introduction to Liquid Helium*, OUP, Oxford (1970).
‡For helium the de Broglie wavelength of translational motion $\lambda \sim h/(mkT)^{1/2}$ is similar to or larger than, the interatomic distance.

14

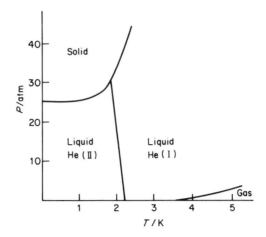

Figure 1.6. The phase diagram for ^{4}He. There is evidence for more than one type of solid but the separate solid phases have not been included in the diagram

isotope ^{4}He. The diagram is unique in three respects. Helium is the only pure substance for which two liquid phases can coexist in equilibrium. These liquids are labelled He(I) and He(II) as shown, and the phase boundary line between them is called the λ line, a name taken from the shape of the curve of heat capacity versus temperature at the phase boundary. He(I) behaves like a normal liquid but He(II) has special properties such as a vanishing viscosity as T approaches zero; by virtue of this property is is called a superfluid.

Helium is the only substance which cannot be solidified at atmospheric pressure however low the temperature. The minimum pressure required to produce the solid is 25 atm. The third unique feature is that there is no boundary line between the solid and the gaseous phases.

At the other extreme of liquid-range temperatures, aluminium typifies the class of liquid metals which is also generally dealt with in specialist texts. We draw attention to the special features of the energies of liquid metals in the next chapter.

Water is such an important liquid that it will be given a chapter to itself in this book so we defer comment at this stage. Benzene is in many respects typical of the wide class of liquids which are formed by non-polar organic molecules; as these are in many respects simpler than water and similar liquids, they will be discussed in the first half of this book.

Chapter 2

Intermolecular Forces

2.1 GENERAL PRINCIPLES

The existence of solid and liquid phases is a manifestation of the attractive forces that act between molecules. It is customary to refer to these attractions as inter-molecular forces† or van der Waals forces with the implication that they are different in degree and in kind from the forces that hold atoms together within a molecule, the so-called intramolecular or valence forces. The extent to which there really are such differences is a question that will be examined later.

We made the point in Chapter 1 that the properties of liquids can be interpreted or predicted at two levels, the microscopic and the macroscopic. It is only in the first of these descriptions that the intermolecular force occurs explicitly. If we wish to understand the properties of liquids in terms of the properties of their component molecules then that understanding must originate in their intermolecular forces.

We begin with a few remarks about force and energy. If two molecules are very far apart there is an attractive force drawing them together. If the molecules are very close to one another there is a repulsive force tending to push them apart. There is a distance, the equilibrium or van der Waals distance, at which the force is zero. It is, as we shall see, often more informative to discuss this interaction between molecules using the inter-molecular energy, rather than the force. The intermolecular energy (also called the potential energy) is defined as the difference between the total energy of the molecules and the sum of the separate molecular energies. The intermolecular force is minus the gradient or slope of the intermolecular energy with respect to distance.

At infinite separations the intermolecular energy is zero. When two

---

†Not all liquids are composed of molecules; liquid argon and mercury are not, for example. We use 'molecular' to encompass atomic species.

16

molecules approach one another the intermolecular energy initially becomes negative; it reaches its lowest value at the van der Waals distance and, if the molecules are brought closer together than this, the energy increases and eventually becomes positive. For a pair of atoms this energy, $U(R)$, is only a function of R, the interatomic distance; typical functions for $U(R)$ and $F(R)$ are shown in Figure 2.1.

For a pair of molecules the intermolecular potential is a function of their relative orientation as well as of R, the distance between their centres. For a general polyatomic molecule the orientation with respect to the vector $\mathbf{R}$ is defined by three angles. However, as the intermolecular potential must be unchanged on rotating both molecules by the same angle around $\mathbf{R}$, the most general potential is a function of R and five angles. The above comments assume that both molecules are rigid structures, and of course they are not; the atoms of a molecule are vibrating even in their lowest energy state (so-called zero-point vibration), and the intermolecular potential will change a little as the atoms move. However, one can generally make a clear distinction between the strong intramolecular energies and the weak intermolecular energies, and in this case the intermolecular potentials can be considered to be averages over the vibrational motions of the two molecules.

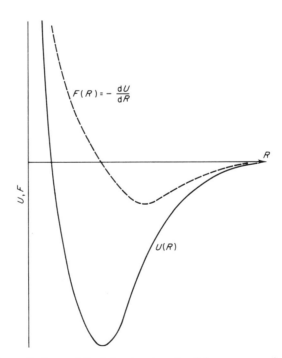

Figure 2.1. Schematic form of the interatomic potential energy as a function of R the atomic separation. The force is given by $-\mathrm{d}U/\mathrm{d}R$

Ab Initio and Empirical Methods

There are two ways in which we can determine the function U: by *ab initio* and by empirical means. *Ab initio* means from first principles. We have to solve the quantum mechanical Schrödinger equation for the electrons for different positions of the nuclei. There is a theorem, due to Born and Oppenheimer, that the energies so obtained give the potential energy function, U, we require. For small atoms, and molecules made up of light atoms (e.g. He, Ne, H_2, H_2O), we can now obtain quite accurate descriptions of U by this means, and for more complex molecules we can at least determine the general functional behaviour of U, such as how it depends on R and on the angles of orientation. With the rapid advance of computer technology there is a parallel rapid advance in our ability to obtain accurate intermolecular potentials by the *ad initio* method.

The word empirical means from experiment. Any experimental observation which can be interpreted theoretically on the basis of the intermolecular potential is capable, in principle, of leading to a potential which is consistent with that observation. However, there are two difficulties with this approach. First, there is not usually a unique potential curve that reproduces a particular experimental observation, although the more data one has available the narrower the range of acceptable potentials. Second, there is usually no simple analytical route for going from the experimental observation to the potential, and one generally has to make the connection by trial-and-error. One assumes a form of the potential and calculates the appropriate experimental result. The difference between this and the observed experimental result is used as a basis for adjusting the form of the potential.

It is almost impossible to achieve a satisfactory potential by trial-and-error if one has no preknowledge of the form of the potential. For that reason the most satisfactory potentials are usualy achieved by semi-empirical means in which some *ab initio* evidence is used to deduce the general features of the potential. To take one example, it can be shown from theory that the interatomic energy for two inert gas atoms at large internuclear distances is proportional to R^{-6}, hence all accurate semi-empirical potentials for these systems are made to satisfy this limiting behaviour.

Many-body Energies

It is quite common to assume that the total intermolecular energy in the solid or liquid states is a sum of the energies of the molecules interacting in pairs and, as we shall see, there is theoretical support for this assumption. In this case we can derive empirical or semi-empirical potentials from experiments that only involve pairwise interactions, and these are mainly experiments on dilute gases. Collisions in a dilute gas are sufficiently infrequent that they nearly all occur between pairs of molecules, with far fewer involving three or more molecules colliding simultaneously. We shall discuss the statistical mechanics of the imperfect gas in Section 3.3 with the particular objecive of seeing how gas properties can be used to derive the pair potential.

In liquids and solids, molecules have simultaneous contacts with several neighbours so that there is no fundamental reason why the bulk properties should depend only on the pair potential. If there are significant contributions to the intermolecular potential from molecules interacting three or more at a time (so-called many-body energies) they will influence the properties of the liquid. Turning the argument the other way round: if the properties of the liquid are used to deduce an empirical pair potential with the assumption that the many-body terms are negligible, then that empirical potential may have little resemblance to the pair potential deduced from gas phase properties or from *ab initio* calculations.

There are reasons, which we shall discuss in the next section, for believing that for liquids composed of inert gas atoms or non-polar molecules the many-body terms can be neglected. For these systems empirical pair potentials deduced from liquid state properties are very similar to the true pair potentials. For liquids composed of polar molecules the many-body terms are probably not negligible, although the situation may well have to be examined for each system of interest. We will give some attention to this point when we come to examine the properties of liquid water in Chapter 8.

Potentials for Metals

There is certainly one class of liquid for which the many-body energies are far from negligible, and that is the liquid metals. The high electrical conductivity of metals shows that electrons are not bound to particular atoms or molecules, but are delocalized over the whole bulk phase. The so-called free-electron model, in which the electrons move in a potential which is constant over the whole phase, is a good first approximation to the electronic states of metals, and for this the total energy depends on the size and shape of the phase, but not on the arrangement of the nuclei.

It would require too large a detour in the development of our subject to cover the theory of the metallic bond, and we must refer the interested reader to specialist texts.† We wish only to make the point that the empirical pair potentials which have been deduced from diffraction data on liquid metals, are quite different from those of, say, the inert gas liquids. Figure 2.2 shows a typical example and we draw attention to the oscillations in the region of large R. These oscillations are a manifestation of the fact that $U(R)$ represents the potential for two atoms in the presence of all other atoms in the system. Such potentials therefore depend on factors such as the

---

†For example, N. H. March, *Liquid Metals*, Pergamon Press, Oxford (1968); T. E. Faber, *Introduction to the Theory of Liquid Metals*, Cambridge U.P., Cambridge (1972); M. Shimoji, *Liquid Metals*, Academic Press, London and New York (1977).

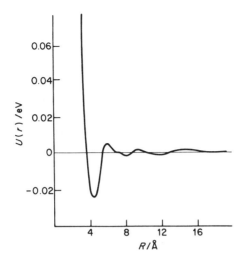

Figure 2.2. The effective pair potential for liquid sodium at 373 K. The potential was deduced by applying the Percus–Yevick equation (mentioned briefly in Section 3.4) to the analysis of the X-ray diffraction data for liquid sodium. (W. S. Howells and J. E. Enderby, *J. Phys.*, C, **5**, 1277 (1972))

density of the liquid, and hence show a slight dependence on temperature.

The term 'effective pair potential' should be used for pair potentials deduced empirically from the properties of any liquid for which the many-body energies are not negligible. This will distinguish such a potential from the true pair potential for two atoms or molecules interacting in the absence of any others.

Inter- and Intramolecular Energies

We return now to the opening statement in this chapter regarding the difference between intermolecular and intramolecular energies. A typical chemical bond has a binding energy of more than $200 \, kJ \, mol^{-1}$; the typical binding energy of neighbouring molecules in a liquid is below $10 \, kJ \, mol^{-1}$. However, there are intermediate cases. The so-called hydrogen bond energy which occurs in water and other hydroxylic solvents may be as large as $25 \, kJ \, mol^{-1}$. Donor–acceptor molecular complexes (e.g. trimethylamine–SO_2) can have binding energies up to $50 \, kJ \, mol^{-1}$.† Although there is not a sharp boundary between the strongest intermolecular bond and the weakest intramolecular bond, the great majority fall clearly into one category or the other and it is therefore convenient to discuss them from different points of view.

---

†The reader should turn to a book on valence theory for a fuller discussion of these. For example, J. N. Murrell, S. F. A. Kettle and J. M. Tedder, *The Chemical Bond*, Wiley, Chichester (1978); R. McWeeny, *Coulson's Valence*, Oxford U.P., Oxford (1980).

Because the intermolecular energy is a small part of the total energy of the system it is possible to utilize a technique of quantum mechanics called the perturbation method to develop the theory of intermolecular energies. Perturbation theory has two pedagogic advantages over the other commonly used technique of quantum mechanics which is the variation method. First, it provides a connection between the intermolecular energy and properties of the interacting molecules, and second, it provides a subdivision of the intermolecular energy into a sum of physically interpretable parts. Its disadvantage is that, in general, it is not as accurate as the variation method nor as easy to implement on a computer. The variation method is almost always the technique which is chosen for calculating intramolecular energies (chemical bond energies), because these are a much larger part of the total energy.

We give in the next section an outline of the perturbation theory of intermolecular energies and direct the reader elsewhere for a fuller treatment.[†]

Short- and Long-range Energies

The most important classification of the various contributions to the intermolecular energy is into short-range and long-range energies. A long-range energy is one that varies at large R as some inverse power of R; say R^{-n}. The Coulombic energy between charges falls into this category having $n = 1$. A short-range energy is one that varies at large R as $\exp(-kR)$, where k is a constant. There is no definite value of R at which short-range energies end and long-range begin. Rather we should think that these contribute to the total energy at all R, but beyond a certain point only the long-range contributions are important. Short-range energies have their origins in the overlap of the wave functions or orbitals of the interacting atoms or molecules. The chemical bond is essentially a short-range energy; the repulsive forces that keep atoms or molecules apart in a solid or liquid are also of this category. Long-range energies do not depend on orbital overlap and, as we shall see in the next section, they can be qualitatively understood from arguments based upon classical electrostatics.

Figure 2.3 shows the interatomic potential for two neon atoms and its division into attractive (long-range) and repulsive (short-range) terms. The van der Waals minimum occurs at the value of R at which the slopes of the two terms are equal in magnitude but opposite in sign. It can be seen that the short-range term is a steeply rising function of R. Indeed, one can say, roughly, that the van der Waals minimum occurs at the distance at which the repulsion starts to be noticeable.

[†]For example, Chapters 2 and 3 in *Rare Gas Solids*, Eds. M. L. Klein and J. A. Venables, Academic Press, London and New York (1976) and Chapter 2 in G. C. Maitland, M. Rigby, E. B. Smith and W. A. Wakeham, *Intermolecular Forces, Their Origin and Determination*, Clarendon Press, Oxford (1981).

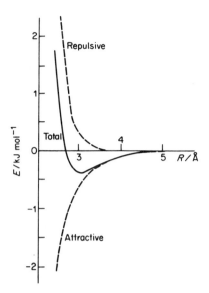

Figure 2.3. Calculated attractive and repulsive energies of Ne–Ne as a function of R

Because the repulsive energies rise steeply when the wave functions of the interacting systems overlap we can introduce as a crude approximation the concept that atoms and molecules have definite sizes and shapes. If atoms were miniature ball bearings the repulsive energy would rise vertically when R equalled the sum of the atomic radii of the two atoms. Although this is an oversimplification it is an important idea because, as we show in Chapter 3, some important features of the liquid state are predicted by models which assume a 'hard-shape' repulsive energy.

In the remainder of this chapter we wish to do two things: to describe the origins and characteristic features of the contributions to the intermolecular potential; and to describe the functional form of semi-empirical potentials that have been commonly used in the theory of the liquid state. For our requirements in the rest of this book the second of these two tasks is the more important, but, we we have already noted, the form of empirical potentials is primarily based on theoretical considerations, and it is therefore appropriate that we tackle these first.

2.2 LONG-RANGE ENERGIES†

Long-range energies can be divided into three types; electrostatic, induction and dispersion energies. To calculate the first of these we need only know

---

†This and the next section will be of interest mainly to those readers who have a strong interest in the theoretical basis of intermolecular forces.

the distribution of charge within the molecules: this is made up of point positive charges from the nuclei and a diffuse negative charge from the electrons. To calculate the other two we need to know how the electrons in a molecule move under the influence of an external electric field. This effect is referred to as the polarization of a molecule; the induction and dispersion energies are both polarization energies. We examine first the electrostatic energy.

Electrostatic Energies

If we know the position of the nuclei and the distribution of the electrons in the molecules, it is possible to calculate the electrostatic energy by summing the nuclear–nuclear repulsions, the electron–electron repulsions and the electron–nuclear attractions. However, such calculations are quite difficult because the electron density is diffuse and an integration over all elemental electron interactions is necessary. Moreover, there is, to a large extent, cancellation of the attractive and repulsive terms making up the total electrostatic energy so that the integrations must be performed with high accuracy.

A much simpler approach is to combine attractive and repulsive terms in such a way that the electrostatic energy can be equated to the energy of interaction of the net charges and multipole moments of the molecules. The resulting expression for the electrostatic energy is known as the multipole expansion. We do not give the formal derivation of this expansion in this book, but note two important points regarding its validity.† First, the expansion is valid only if there is negligible overlap of the electron clouds of the two molecules, and second, for the expansion to be convergent in the first few terms, the distance between the centres of the two molecules must be large compared with the linear dimensions of the two molecules. Broadly speaking, it is only when the two molecules are far apart that the multipole expansion is valid and rapidly convergent.

At short distances, say less than van der Waals' separations, the multipole expansion will provide only a qualitative guide to the electrostatic energy, and to obtain an accurate value it is usually necessary to sum the individual Coulombic energies, as already described. A useful approach in this case is to represent the total charge distribution by a set of charges and dipoles centred on each atom; this is called the distributed multipole method.

Electric Moments

In many molecules, there is a distribution of electric charge; to give an example, a central positively-charged atom may be attached to a number of

---

†For a derivation see A. D. Buckingham, *Quart. Revs.*, **13**, 183 (1959); and Chapter 1, *Intermolecular Interactions from Diatomics to Biopolymers*, Ed. B. Pullman, Wiley–Interscience, Chichester (1978).

negatively-charged atoms such that there is overall electrical neutrality. To describe quantitatively the interaction between such a molecule and another charge or group of charges, several important mathematical functions are employed; known as the moments of the distribution, they involve the magnitudes of the various charges and the distances separating them. When there is effectively a dipole, i.e. two equal separated charges $+\delta$ and $-\delta$, the relevant function is called the first moment of the distribution, and it is simply the product of the magnitude of the charge and the distance of separation r; it is more generally known as the dipole moment δr, and it is considered to be directed along a line from the positive to the negative charge. The traditional unit for molecular dipole moments is the debye†. (A more general definition of the dipole moment is given in the appendix to this chapter.)

A molecule such as CO_2 has a zero dipole moment because the two dipoles associated with the carbon–oxygen bonds exactly cancel out; nevertheless it is capable of interacting with a charge, a dipole or a multipole in a way which is characterized by the second moment of the charge distribution or quadrupole moment (symbol Θ), which represents the effective behaviour of the set of separated charges, $+2\delta$ on the carbon atom and $-\delta$ on each of the oxygen atoms.

Molecules of higher symmetry may have zero dipole and quadrupole moments but still interact with external charges by virtue of a non-zero third (octopole) or higher moments of the charge distribution. Thus, CH_4 has a significant third moment and SF_6 a significant fourth moment, even though all the lower moments are equal to zero in both cases. (If the distribution of one type of charge falls on a uniform spherical shell around the exactly-compensating charge at the centre, as with a neon atom, all the moments are equal to zero, but this cannot happen in a molecule.) A pole (dipole quadrupole, etc.) of order a (for a dipole $= 1$, quadrupole $= 2$, octopole $= 3$, etc.) interacts with a pole of order b separated by a distance R with a potential energy proportional to $R^{-(a+b+1)}$.

The multipole expansion for the electrostatic energy is an expansion in powers of $1/R$, where R is the distance between the centres (usually taken as the centres of mass) of the two molecules. The first term is the Coulomb energy between the net charges q_a, q_b of the two molecules‡

$$q_a q_b / 4\pi\varepsilon_0 R. \tag{2.1}$$

The second term is the interaction between the charge on one molecule and the dipole moment μ of the other. Because the dipole is a vector quantity the energy depends on the orientation of the dipoles; in terms of the angular variables defined in Figure 2.4 it is

$$(q_b \mu_a \cos\theta_a - q_a \mu_b \cos\theta_b)/4\pi\varepsilon_0 R^2. \tag{2.2}$$

---

†1 debye $= 10^{-18}$ e.s.u. $= 3.3356 \times 10^{-30}$ C m.
‡$4\pi\varepsilon_0 = 1.11 \times 10^{-10}$ C V m^{-1} = 1 e.s.u.

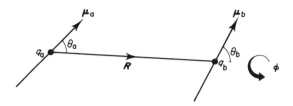

Figure 2.4. Coordinates for the electrostatic energy between charges and dipole moments

Equation 2.2 is more compactly written in vector notation

$$(q_b\boldsymbol{\mu}_a \cdot \boldsymbol{R} - q_a\boldsymbol{\mu}_b \cdot \boldsymbol{R})/4\pi\varepsilon_0 R^3, \tag{2.3}$$

where $\boldsymbol{R}$ is the vector distance of B from A.

The first non-zero contribution to the multipole expansion for neutral but dipolar molecules is the dipole–dipole energy

$$-\mu_a\mu_b(2\cos\theta_a\cos\theta_b - \sin\theta_a\sin\theta_b\cos\phi)/4\pi\varepsilon_0 R^3, \tag{2.4}$$

which in vector form is

$$\left(\frac{\boldsymbol{\mu}_a \cdot \boldsymbol{\mu}_b}{R^3} - 3\frac{(\boldsymbol{\mu}_a \cdot \boldsymbol{R})(\boldsymbol{\mu}_b \cdot \boldsymbol{R})}{R^5}\right)\Big/4\pi\varepsilon_0. \tag{2.5}$$

The most favourable (most negative energy) orientations of two dipoles are colinear head-to-tail $\theta_a = \theta_b = 0$, or side-by-side (antiparallel) $\theta_a = \theta_b = \pi/2$, $\phi = \pi$.

The full expressions for the higher terms in the multipole expansion are rather complicated and will not be given here.† For charged systems there is another term like 2.7 which is proportional to R^{-3} due to the interaction of the charge on one molecule with the quadrupole of the other. The terms in R^{-4} are charge–octopole and dipole–quadrupole, and so on. A result that summarizes the whole series is that the mth moment of one molecule interacting with the nth moment of another gives an energy proportional to $R^{-(m+n+1)}$. The charge can be included in this formula by considering it to be the zeroth moment.

Table 2.1 gives two examples of the rate of convergence of the multipole expansion. Athough dipole and quadrupole moments are often known from experiment this is not generally the case for higher moments. However, quite accurate values for these can be obtained for small molecules from quantum mechanical calculations, and such values have been used to give the results in Table 2.1. The values of R chosen for these calculations are roughly the van der Waals distances for the dimers. At shorter distances the overlap of the electron clouds of the two molecules makes the multipole expansion invalid.

---

†A. D. Buckingham, see reference on p. 22.

Table 2.1. Electrostatic energies obtained from the multipole expansion with the multipoles terminated at different order

	LiH ($R = 14a_0$)		HF ($R = 6a_0$)	
	$\rightarrow\rightarrow$	$\uparrow\downarrow$	$\rightarrow\rightarrow$	$\uparrow\downarrow$
Dipoles	-40.6	-20.3	-54	-27
Dip. + quad.	-39.4	-19.8	-30	-18
Dip. + quad. + oct.	-41.7	-19.0	-56	-12
Est. series limit	-42	-19	-49	-15

Distances and energies in atomic units, $a_0 = 0.5295$ Å, unit of energy 10^{-4} $E_h = 0.2626$ kJ mol^{-1}. The diatomic bond lengths are $3.02a_0$ and $1.73a_0$ for LiH and HF respectively J. T. Bröbjer and J. N. Murrell, *Chem. Phys. Letters*, **77**, 601 (1981).

The results of Table 2.1 show that the multipole expansion converges rapidly for the LiH dimer but slowly for the HF dimer. This is due not only to the different values of R in the two cases but also because the higher multipole moments are larger, relative to the dipole, for HF than for LiH. These results show that care is needed in calculating electrostatic energies by the multipole expansion. The distributed multipole method, mentioned earlier, is generally more reliable.

For molecules which have no dipole moments but large quadrupole moments such as CO_2 ($\Theta = -14 \times 10^{-40}$ C m^2 $= -4.2 \times 10^{-26}$ e.s.u.), the quadrupole–quadrupole contribution to the electrostatic energy can be quite large. For example, side-by-side CO_2 molecules separated by 4 Å will have a positive energy from this source equal to

$$\frac{9\Theta^2}{16\pi\varepsilon_0 R^5} = 3.9 \times 10^{-21} \text{ J } (= 2.3 \text{ kJ mol}^{-1}). \tag{2.6}$$

Polarization Energies

We turn now to the polarization energies which are due to changes in the electron distribution within the molecules when they come within each other's influence. The electric fields arising from the permanent moments of one molecule will induce moments in a neighbour, and the interaction between permanent and induced moments will lead to an attraction between the two: this is called the induction energy. The type of electron displacement is illustrated for a simple case in Figure 2.5. The induction energy can be large for interacting ions or molecules with large dipole moments, but it will be small for nearly spherical molecules like CH_4 and SF_6, because these have no low-order moments, and the electric fields in their vicinity are small. It will be zero for inert gas atoms as these have no moments of any order.

Although an inert gas atom has no permanent moments it can be

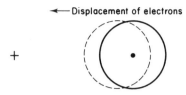

Figure 2.5. The polarization of an atom by a neighbouring positive charge

considered to possess time-varying (instantaneous) moments due to fluctuations in its electron distribution. These instantaneous moments will induce moments in a neighbouring molecule and there will be an energy of interaction between the two. Although the instantaneous moments average over time to zero, the polarization energy that is associated with them does not average to zero. It is this energy that is called the dispersion energy. A rigorous theory of dispersion energies can only be obtained by quantum mechanics. This theory was first derived by F. London, and hence dispersion energies are often referred to as London energies.

The extent to which the electron density in an atom or molecule is polarized by an external field is measured by a quantity α called the polarizability, which is defined as the dipole moment induced by a unit applied field $\mathscr{E}$.†

$$\boldsymbol{\mu}_{\text{ind}} = \boldsymbol{\alpha} \cdot \mathscr{E}. \tag{2.7}$$

Both $\boldsymbol{\mu}_{\text{ind}}$ and $\mathscr{E}$ are vector quantities so that $\boldsymbol{\alpha}$ is a tensor having components α_{xx}, α_{xy} etc. For spherical systems $\alpha_{xx} = \alpha_{yy} = \alpha_{zz} = \alpha$ and $\alpha_{xy} = \alpha_{yz} = \alpha_{zx} = 0$. For non-spherical systems the average polarizability, a scalar quantity, is defined by

$$\alpha = \tfrac{1}{3}(\alpha_{xx} + \alpha_{yy} + \alpha_{zz}). \tag{2.8}$$

The quantity α/ε_0 has the dimension of volume, and its value is roughly proportional to the molecular volume. For example, for N_2 α is approximately $12a_0^3\varepsilon_0$ and for benzene approximately $70a_0^3\varepsilon_0$. The polarizability of a molecule like benzene is very anisotropic as the electrons are much more mobile for movement parallel to the plane of the molecule than for movement perpendicular to the plane: this is due to the delocalized π electrons of the conjugated double bonds.

Induction Energies

The induction energies can be calculated by substituting the expression for the induced moment (2.7) for one of the dipole moments which occur in expressions 2.2 or 2.4. For example, if molecule A has a charge q_a this will

---

†Some texts include a factor $4\pi\varepsilon_0$ in this expression, but we prefer the convention in which this is subsumed within α.

give a field at B proportional to $q_a R^{-2}$, and the induced dipole moment will be proportional to $q_a \alpha_b R^{-2}$. The interaction energy of this charge with the induced moment is proportional to $q_a^2 \alpha_b R^{-4}$. The full expression with appropriate factors is (assuming an isotropic polarizability)

$$E_{ind} = -\tfrac{1}{2} q_a^2 \alpha_b / (4\pi\varepsilon_0)^2 R^4. \tag{2.9}$$

Note that induction energies are always negative because the charge or permanent moment always induces a moment in the sense that their interaction is attractive. This is illustrated in Figure 2.6, which shows the direction of the induced dipole arising from the field of a permanent dipole for various orientations of two molecules. The induction energy for a molecule with a polarizability α_b in the field of a dipole moment μ_a, when averaged over all orientations, is given by the following expression

$$E_{ind} = \frac{\mu_a^2 \alpha_b}{(4\pi\varepsilon_0)^2 R^6}. \tag{2.10}$$

Dispersion Energies

Figure 2.6 can also be used to show that the dispersion energy is always negative. If the permanent dipole is replaced by a temporary dipole then the induced dipoles will always be in such a direction that they are attracted to the temporary dipole. For example, if the temporary dipole is reversed, all the induced dipoles will reverse. Thus, even if the average of the temporary dipole is zero, as it must be for a molecule without a permanent dipole moment, the average of the interaction energies with which it is associated will not be zero.

The electrostatic field at a distance R from a dipole is proportional to R^{-3}, hence the induced moments will, from equation 2.7, also be proportional to R^{-3}. As dipole–dipole energies are proportional to R^{-3} (2.4),

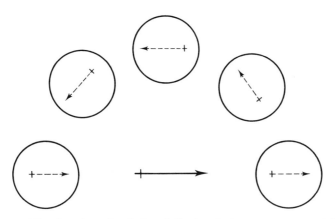

Figure 2.6. The direction of an induced dipole relative to a permanent dipole

dipole–induced dipole energies will be proportional to R^{-6}. It is therefore customary to write the dispersion energy as

$$E_{disp} = -C_6/R^6, \tag{2.11}$$

where the coefficients C_6 can be deduced empirically or by calculation. For atoms or nearly spherical molecules C_6 is independent of the relative orientation of the systems. For other molecules C_6 will be angle-dependent but in most cases only spherical averages are known. Table 2.2 gives some typical values for C_6. These are quoted in atomic units† so that to obtain the energy in atomic units one must insert in equation 2.11 the appropriate value of R in atomic units.

Expression 2.11 is the first term in an expansion in $1/R$ for the dispersion energy, with higher terms (C_8/R^8 etc.) arising from the interaction of induced multipoles of higher order (quadrupoles, etc). The values of C_8 and C_{10} are known for simple systems like the inert gas atoms, but these higher terms are usually less than 10% of the leading term at van der Waals distances.

An approximate relationship between the dispersion energy and the polarizabilities of the interacting systems was derived by London using perturbation theory

$$E_{disp} = -\frac{3I_aI_b}{2(I_a + I_b)} \cdot \frac{\alpha_a\alpha_b}{(4\pi\varepsilon_0)^2R^6}. \tag{2.12}$$

I_a and I_b are ionization potentials of the two systems. For two identical atoms or molecules this formula becomes

$$E_{disp} = -\frac{3\alpha^2I}{4(4\pi\varepsilon_0)^2R^6}. \tag{2.13}$$

It can be seen from expressions 2.10 and 2.12 that for dipolar molecules the induction and dispersion energies are both proportional to R^{-6}. Using the London formula for the dispersion energy, 2.13, the ratio of the

Table 2.2. Values of the dispersion energy coefficient $C_6/E_h a_0^6$ defined by expression 2.11

	H	He	Ne	H$_2$	N$_2$	CH$_4$
H	6.50	2.83	5.6	8.7	21	30
He	2.83	1.47	3.0	4.1	10	14
Ne	5.6	3.0	6.3	8.2	21	29
H$_2$	8.7	4.1	8.2	13	30	43
N$_2$	21	10	21	30	73	100
CH$_4$	30	14	29	43	100	150

Values are from A. Dalgarno, *Adv. Chem. Phys.*, **12**, 143 (1967). $E_h a_0^6 = 9.573 \times 10^{-80}$ J m^6.

---

†$a_0 = 0.5292$ Å, E_h (the hartree) $= 2626$ kJ mol^{-1}.

Table 2.3. Data relevant to the induction and dispersion energies between HCl molecules

$\mu = 1.08$ D $= 3.60 \times 10^{-30}$ C m
$\alpha = 2.6 \times 4\pi\varepsilon_0 \text{Å}^3 = 2.9 \times 10^{-40}$ J^{-1} C^2 m^{-1}
$I = 12.75$ eV $= 2.04 \times 10^{-18}$ J

$$\frac{\mu^2}{\alpha I} = 0.022$$

induction energy to the dispersion energy is approximately $\mu^2/\alpha I$. The size of this factor in a typical case is shown by the data in Table 2.3. The induction energy is only 2% of the dispersion energy. As polarizabilities are roughly proportional to molecular volumes it is, in general, only for small molecules with large dipole moments that the induction energy is important. For ionic systems the induction energy (an R^{-4} term as shown by equation 2.9) may also be large, but in this case the electrostatic energy is by far the most important contribution to the long-range energy.

Pair Additivity

The final point to which we give attention in this section is the pair additivity of the long-range energies. For example, if we have three molecules A, B and C in mutual proximity then we wish to know whether the total intermolecular energy is equal to the sum of A–B, B–C and A–C pair energies. Any deviation can be attributed to a three-body energy.

The electrostatic energy is exactly pair-additive: for example, the electrostatic energy for three point charges is the sum of the three Coulombic energies for the charges interacting in pairs. It can be shown by perturbation theory that the dispersion energy is also exactly pair-additive in the R^{-6} term (2.11), with non-additive contributions only appearing in higher powers of $1/R$.†

The induction energy is not pair-additive as can easily be seen by qualitative argument. Figure 2.7 shows the displacement of the electrons of an atom towards a neighbouring positive charge. If a second positive charge is placed at the same distance on the other side of the atom, the fields from the two positive charges will cancel and there will be no displacement of the centre of the electron cloud, only a symmetrical distortion as shown. There is therefore a charge-induced dipole energy with one charge present but no such term with two charges present. The net induction energy is therefore not the sum of the induction energies arising from one charge being there without the other. It is because of the possible importance of induction energies in determining the properties of polar liquids that one must be cautious in assuming pair-additive potentials for such systems.

---

†For a proof of this see Chapters 2 and 3 in *Rare Gas Solids*, Eds. M. L. Klein and J. A. Venables, Academic Press, London and New York (1976).

30

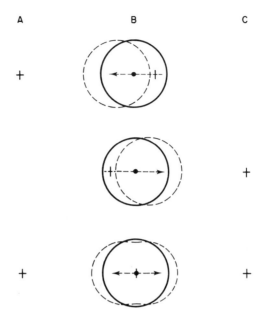

Figure 2.7   An illustration of the non-pair additivity of induction energies

2.3   SHORT-RANGE ENERGIES

Short-range energies arise from the overlap of electron densities of the interacting molecules. If the two species have unpaired electrons (e.g., H, CH_3) then by pairing the electron spins so that the net spin angular momentum is zero, an attractive short-range energy is obtained. This is the origin of intramolecular (valence) energies, i.e. of the covalent bond. If the two species have no unpaired electrons (e.g., He, CH_4) they are said to be closed-shell systems and the short-range energy is repulsive. A qualitative explanation for this repulsion is that when orbitals containing electrons of the same spin overlap there is an approaching violation of the Pauli exclusion principle that two electrons cannot have the same orbital wave function and the same spin; a repulsive force is associated with this violation

The short-range repulsion is often referred to as exchange repulsion. The reason for this is that when the wave functions of the two molecules overlap we can no longer identify electrons as belonging to one or the other molecule and the total wave function must reflect this fact by allowing for exchange between the two.

If two molecules A and B have orbitals with wave functions ϕ_a and ϕ_b, respectively, the overlap of these orbitals is measured by the overlap integral

$$S = \int \phi_a \phi_b \, dv, \tag{2.14}$$

where the integration is over all space. This integral measures the amount of space which is common to both wave functions. If A and B are far apart the orbitals ϕ_a and ϕ_b will not overlap significantly and S will be effectively zero.

A common feature of the wave functions of atomic and molecular orbitals is that they decay exponentially to zero at infinity. It follows that the overlap integral also decays exponentially to zero as the distance between the two molecules goes to infinity. To give a specific example, the overlap integral between two hydrogen 1 s atomic orbitals has the form

$$S = (1 + R + R^2/3) \exp(-R). \qquad (2.15)$$

The exchange energy involves the product of two orbital overlap functions, $\phi_a \phi_b$, so that the exchange energy is approximately proportional to S^2. As the exchange energy is rather difficult to calculate but the overlap integral easy, this has provided a useful route to the estimation of exchange energies. The following approximate relationship was suggested by Mulliken

$$E_{ex} = kS^2/R \qquad (2.16)$$

where k is a constant the value of which depends on the type of orbitals involved. A more general approach is to assume that the exchange energy is of the form

$$E_{ex} = A \exp(-bR) \qquad (2.17)$$

where A and b are parameters the values of which are to be determined empirically. Expression 2.17 is known as the Born–Mayer potential.

The exchange energy can generally be taken as pair-additive. To obtain a significant three-body exchange energy it is necessary for there to be some region of space where there is simultaneously an overlap of the wave functions of three molecules, and this would only occur if the pair overlaps were exceptionally large. In other words it would only occur in very repulsive situations which are not of any significance for the liquid state.

The Hydrogen Bond

Amongst those systems where the intermolecular energy is exceptionally large the hydrogen bond is by far the most important for the liquid state. In Chapter 8 we will examine in detail the intermolecular potential for water molecules, but a few more general comments are in order at this point.

The hydrogen bond is a structure in which a hydrogen atom is partly bonded to two electronegative atoms, most commonly nitrogen, oxygen or a halogen atom, the three atoms being usually colinear. In most cases the hydrogen is more strongly bound to one atom (A) than the other (B), the A–H bond being almost the same length as in the uncomplexed system A–H, and the other bond B–H being longer than a normal chemical bond.

Table 2.4. Characteristic properties of the perturbation contributions to intermolecular energies

Type	Range	Attractive or repulsive	Pairwise additive
Electrostatic	Long	Either	Yes
Exchange	Short	Repulsive	Yes
Dispersion	Long	Attractive	Yes
Induction	Long	Attractive	No
Exchange polarization or charge transfer	Short	Attractive	No

In a few strong complexes the hydrogen is symmetrically placed between A and B.

Recent *ab initio* calculations on hydrogen-bonded complexes have confirmed the early view that the main contribution to the binding energy comes from the electrostatic energy between the dipolar A–H bond and a partial negative charge on the electronegative atom B. For the stronger complexes, however, there is a significant contribution to the energy from a valence-type interaction arising from the overlap of orbitals of A–H with those of B. This interaction leads to a partial transfer of electrons from B to the A–H bond.†

A negative (stabilizing) energy arising from the overlap of orbitals of closed-shell systems (as distinct from the exchange repulsion already discussed) is commonly explained by the Mulliken theory of charge-transfer or donor–acceptor complexes. We again refer the reader elsewhere for details.‡ A point that should be made, however, is that the relationship of charge-transfer to exchange energies has some similarity to the relationship of induction and dispersion to electrostatic energies. Charge-transfer energies can be thought of as involving exchange and polarization, and the alternative description exchange-polarization is often used. Like induction energies, charge-transfer energies are not pair-additive.

Table 2.4 summarizes some of the conclusions that have been reached in the last two sections.

2.4 EMPIRICAL AND SEMI-EMPIRICAL POTENTIALS

Although it is possible to predict from quantum mechanics the general functional form of the intermolecular potential, it is only for relatively simple atoms or molecules that we can obtain, in this way, potentials which

---

†For a more detailed discussion see L.C. Allen, *J. Amer. Chem. Soc.* **97**, 6921 (1975).
‡J. N. Murrell, S. F. A. Kettle and J. M. Tedder, *The Chemical Bond*, Ch. 15, Wiley, Chichester (1978).

are sufficiently accurate to provide a very good fit to the bulk properties of gases, liquids and solids. As explained in the first section of this chapter, a more successful route for obtaining accurate potentials is to combine some theoretical knowledge on the form of a potential with experimental information on the values of parameters which occur in the potential. This is called the semi-empirical method.†

Even if a potential is known not to be an accurate representation of the true potential, an analysis of the type of bulk properties that it can give may be revealing. For example, the simplest pair potential for atoms is the hard-sphere function. If an atom is modelled by a hard-sphere of diameter σ the potential will be zero for $R > \sigma$ and infinite for $R < \sigma$ with a discontinuity at $R = \sigma$. This function is illustrated in Figure 2.8(a). It can be shown that the hard-sphere potential, when introduced into statistical theories, reproduces qualitatively many of the properties of a fluid: the thermal conductivity and viscosity are two examples, and a value of σ can be chosen to give the best fit to such properties. It also leads to a bulk phase which has a solid–liquid phase change. The important inference from such an analysis is not that the hard-sphere potential is an accurate representation of the potential, but that the particular properties mentioned are primarily dependent on the repulsive part of the potential, and for such properties the van der Waals attractive well is much less important.

To obtain a liquid–gas phase change it is necessary to use a potential which has a minimum. The simplest potential which possesses this can be obtained by combining a square-well attraction with the hard-sphere repulsion as shown in Figure 2.8(b). Unfortunately this potential has not proved to be mathematically very easy to use in theories of fluids and, as it is a long way from being a realistic potential, not a great deal of attention has been given to it in the scientific literature.

The potential which has been the basis of most analytical studies of atomic fluids and is at the same time a reasonably close representation of real potentials has the form

$$U = 4\varepsilon\left[\left(\frac{\sigma}{R}\right)^{12} - \left(\frac{\sigma}{R}\right)^{6}\right] \tag{2.18}$$

It is called the Lennard-Jones or L-J potential after its originator. It is also often called the 6–12 potential being one member of the family

$$U = A\left[\left(\frac{\sigma}{R}\right)^{m} - \left(\frac{\sigma}{R}\right)^{n}\right], \qquad m > n, \tag{2.19}$$

where m and n are integers. These potentials are zero at $R = \infty$ and $R = \sigma$, and have a minimum at

---

†A very detailed account of the method of determining intermolecular energies from experimental data is contained in *Advances in Chemical Physics, Volume 12, Intermolecular Forces*. Ed. J. O. Hirschfelder, Interscience, London and New York (1967), also G. C. Maitland *et al*, see reference on p. 20.

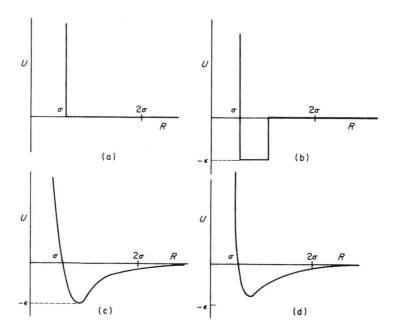

Figure 2.8. A comparison of simple empirical potentials: (a) hard-sphere, (b) square-well, (c) Lennard-Jones (equation 2.18), (d) (exp–6), $[U = 9 \times 10^4 \varepsilon \exp(-10R/\sigma) - 4\varepsilon(\sigma/R)^6]$

$$R_{\min} = \sigma \left(\frac{m}{n} \right)^{1/(m-n)}, \tag{2.20}$$

of depth

$$U_{\min} = A \left[\left(\frac{n}{m} \right)^{m/(m-n)} - \left(\frac{n}{m} \right)^{n/(m-n)} \right], \tag{2.21}$$

For the L-J potential $R_{\min} = 2^{1/6}\sigma$, $U_{\min} = -\varepsilon$ as indicated in Figure 2.8(c).

At large R the L-J potential is asymptotic to an R^{-6} curve and hence it has the correct form to reproduce the long-range dispersion energy between closed-shell atoms and molecules. For $R < \sigma$ the first term in the potential dominates and making this proportional to R^{-12} ensures a rapid rise in the repulsive branch of the curve. There is no physical significance in taking $m = 12$, but it is mathematically advantageous to have $m = 2n$ for the evaluation of certain integrals that enter the calculation of fluid properties.

There is now general agreement that one gets closer to exact interatomic potentials by taking the repulsive part of the potential to be an exponential; the theoretical arguments for this were given in the last section. The most commonly employed simple function is

$$U = A \exp(-kR) - C_6/R^6, \tag{2.22}$$

which is usually called the (exp − 6) potential. This function has the merit that both the long-range attraction and short-range repulsion are functions that are supported by theoretical analysis. A typical function is shown in Figure 2.8(d).

All the functions illustrated in Figure 2.8 have so few constant parameters (from one to three) that the values of these parameters can be optimized by fitting theoretical predictions to experimental data and, as explained in the first part of this chapter, these data are usually obtained from the properties of gases. However, no simple analytical function, either those we have described or others, gives an exact representation of the data. To obtain agreement between theory and experiment at the highest possible level most workers have employed different functions for different ranges of R. It is then necessary to impose the condition that U shall be a smooth function over the complete range of R, and this places further restrictions on the parameters. Very accurate pair potentials for the inert gases have been obtained in this way.†

The current situation for the potentials between molecules is one of active research, but there is no widely held view on what are the best semi-empirical functions to employ. One popular approach is to assume that the potential is a sum of atom–atom pair potentials using functions of the type illustrated in Figure 2.8: a sum of L-J potentials has been most commonly used. For example, the intermolecular potential between two N_2 molecules would be the sum of four N–N pair potentials. For polar molecules it is common to add to such terms an electrostatic energy in the form of the multipole expansion. The potential between two HCl molecules would therefore be the sum of H–H, Cl–Cl and H–Cl pair potentials, plus electrostatic terms such as expression 2.5 obtained from dipole and quadrupole moments.

The most complicated fluid which has been extensively subjected to theoretical analysis is water and the potentials employed for such studies will be described in Chapter 8.

2.5 LONG-RANGE MACROSCOPIC ATTRACTION

The existence of long-range attractive forces between molecules and macroscopic particles or between two macroscopic particles, is very significant for an understanding of several topics that are dealt with in the last three chapters of this book: colloid stability, wetting phenomena, membrane properties, etc.‡ To account for these forces it is necessary to extend the theory of molecule–molecule interactions in two ways. First, it is necessary

---

†For example, an accurate Ar–Ar potential has been published by J. M. Parson, P. E. Siska and Y. T. Lee, *J. Chem. Phys.*, **56**, 1511 (1972).
‡P. Richmond, *Colloid Science*, Specialist Periodical Report, Chem. Soc. Vol. 2, (1975); A. W. Adamson, *Physical Chemistry of Surfaces*, 3rd Edn, Wiley, Chichester (1977).

to sum the forces from all the molecules that make up the macroscopic particles, and second, a correction has to be made to the long-range forces when they operate over macroscopic distances (typically greater than 500 Å).

We have seen (2.11) that the long-range dispersion energy between two molecules has the form $-C_6/R^6$; C_6 is of the order of $10^{-78}\,\mathrm{J\,m^6\,atom^{-1}}$. The sum of such contributions for all molecules in a macroscopic particle will depend on the shape of the particle. The most important cases are when the particle has a planar face, which can be considered as infinite in area, and is of infinite thickness (this is called the semi-infinite solid), and when the particles are spheres of uniform radii. In these cases the total dispersion energy can be obtained by integrating expression 2.11 between certain limits. The following are the most important results.

(1) For a molecule at a distance R from the surface of a semi-infinite solid the energy is

$$E = - \frac{\pi n C_6}{6R^3} \tag{2.23}$$

where n is the number density of molecules (number per $\mathrm{m^3}$) in the solid.

(2) For two semi-infinite solids whose faces are parallel and separated by a distance R, the energy per unit area of face is given by

$$E = - \frac{H}{12\pi R^2}, \tag{2.24}$$

where H is known as the Hamaker constant, which for this system is given by

$$H = \pi^2 n^2 C_6. \tag{2.25}$$

(3) For two spheres of radius a whose centres are d apart, the energy is

$$E = - \frac{H}{6} \left\{ \frac{2}{D^2 - 4} + \frac{2}{D^2} + \ln\left(\frac{D^2 - 4}{D^2}\right) \right\}, \tag{2.26}$$

where $D = d/a$. If the spheres are large compared with the shortest distance between the surfaces $(d \approx 2a)$ then the Hamaker constant is also given by expression 2.25, and this is the usual assumption.

The most important point to make about expressions 2.23, 2.24 and 2.26 is that the energy depends on a smaller inverse power of the separation than does the molecule–molecule energy. As a result these energies are significant over much larger distances than the molecule–molecule energy.

In Section 2.2 we gave the following qualitative explanation of dispersion forces: a temporary dipole on one molecule induces a dipole on a neighbouring molecule and there will be an attraction between these two dipoles. In standard quantum mechanics this interaction is considered to be instantaneous, but in quantum electrodynamics, a more advanced theory, it is recognized that it takes time for the interaction to be passed between the molecules by an electromagnetic radiation field. For large separations this results in a difference in phase between the temporary and induced dipoles, and hence some reduction in the dispersion force. Such a reduction was first postulated by Verwey to explain the apparently small value of dispersion energies between colloidal particles at large distances. The first detailed theory using quantum electrodynamics, was advanced by Casimir and Polder in 1948.

If the temporary dipoles in a molecule fluctuate with a frequency v there will be an associated radiation of wavelength $\lambda = c/v$. If the distance between the molecules is comparable with or greater than λ then the phase difference of the temporary and induced dipoles will be significant. Typical wavelengths are those associated with electronic transitions in the molecules, 10^3–10^4 Å (10^2–10^3 nm), and it is at such distances or longer that one refers to the forces as being very-long-range or retarded. Casimir and Polder showed that in this region the dipole–dipole dispersion energy varied as R^{-7} rather than R^{-6}. The retarded forms of the macroscopic interactions given in 2.23, 2.24 and 2.26 also increase by one power of R in the denominator.

The London theory of dispersion energies relates the C_6 coefficient to polarizabilities of the molecules (2.12). In 1956 Lifshitz showed that it was possible to relate macroscopic dispersion energies directly to macroscopic properties of the interacting particles and the intervening medium. The relevant property is the dielectric constant (relative permittivity), which, as we show in Chapter 8, is the macroscopic analogue of polarizability. The Hamaker constant that appears in 2.24 and 2.26 can be determined from a knowledge of the dielectric constant as a function of frequency, although the method is quite complicated and we give no details here.†

Hamaker constants for single materials are typically of the order of 10^{-19} J and some values calculated by the macroscopic approach are given in Table 2.5. The presence of a liquid medium between the particles, rather than a vacuum or air, lowers appreciably the dispersion energy, and it also shortens the distance at which retardation sets in. For a system of particles 1 and 2 separated by an intervening medium 3 the effective Hamaker constant for the particles is given by

$$H_{12,3} = H_{12} + H_{33} - H_{13} - H_{23}. \tag{2.27}$$

Note that if the medium (3) has the same dielectric properties as 1 and 2, then there is effectively no interaction between the particles.

Finally, it should be mentioned that there has been much interest in the direct experimental measurement of the attractive forces between macro-scopic systems at sub-micron distances; notable in this area is the work of

---

†P. Richmond, see reference on p. 35. D. B. Hough and L. R. White, *Adv. Colloid Interf. Sci.*, **14**, 3 (1980).

Table 2.5. Values of the Hamaker constant
for different media, calculated from macroscopic theory

	$H/10^{-20}$ J
Water	3.0–6.1
Ionic crystals	5.8–11.8
Metals	22
Silica	8.6
Quartz	8.0–8.8
Hydrocarbons	6.3
Polystyrene	5.6–6.4

Where a range of values is given this reflects different
methods of calculation. J. Visser, *Surface and Colloid
Science*, Ed. E. Matijević, Vol. 8, Wiley–Interscience,
Chichester (1976).

Tabor and co-workers on mica.† There has been a broad measure of
agreement with modern theory down to separations of about 1 nm.

APPENDIX 2

The net charge of a molecular species is obtained by summing over the nuclear
charges eZ_α and integrating over the electron distribution. If $\rho(r)$ is the electron
probability density then

$$q = e\left(\sum_\alpha Z_\alpha - \int \rho(r)\,dv\right). \tag{A2.1}$$

The first moment is the dipole moment which is a vector quantity whose x
component is, for example, given by

$$\mu_x = e\left(\sum_\alpha Z_\alpha x_\alpha - \int \rho(r).x.dv\right). \tag{A2.2}$$

The second moment, or quadrupole, is a tensor with components Θ_{xx}, Θ_{yy}, Θ_{zz},
$\Theta_{xy} = \Theta_{yz}$, etc., given by the expressions

$$\Theta_{xx} = \frac{e}{2}\left(\sum_\alpha Z_\alpha(3x_\alpha^2 - r_\alpha^2) - \int \rho(r)(3x^2 - r^2)\,dr\right), \tag{A2.3}$$

$$\Theta_{xy} = \frac{3}{2}e\left(\sum_\alpha Z_\alpha x_\alpha y_\alpha - \int \rho(r).xy\,dv\right) \text{ etc.} \tag{A2.4}$$

For a molecule with a three-fold or higher symmetry axis (e.g. CH_3F or a linear
molecule) there is only one independent component of the quadrupole tensor. If z is
taken as the symmetry axis, then $\Theta_{zz} = -2\Theta_{xx} = -2\Theta_{yy}$, $\Theta_{xy} = \Theta_{yz} = \Theta_{zx} = 0$.

†J. M. Israelachvili and D. Tabor, *Proc. Roy. Soc.*, **A331**, 19 (1971).

Only the first non-zero moment of a molecule is independent of the origin of the coordinate system. For example, suppose the origin is displaced in the x direction by a so that $x' = x - a$, then expression A2.2 becomes

$$\mu'_x = e\left(\sum_\alpha Z_\alpha x'_\alpha - \int \rho(r).x' \, dv\right)$$

$$= e\left(\sum_\alpha Z_\alpha (x - a) - \int \rho(r)(x - a) \, dv\right)$$

$$= \mu_x - ae\left(\sum_\alpha Z_\alpha - \int \rho(r) \, dv\right)$$

$$= \mu_x - aq. \tag{A2.5}$$

Hence for a neutral molecule ($q = 0$), $\mu'_x = \mu_x$, but for a charged species this is not the case. Likewise for a dipolar molecule the quadrupole is dependent on origin.

In the multipole expansion of the electrostatic energy R is the distance between the origins of the multipole moments of the two molecules, and hence the individual terms in the multipole expansion are origin-dependent both in their numerators (the multipoles) and their denominators (powers of R). However, the total electrostatic energy is independent of origin. It is usual to choose the centre of mass of a molecule for the multipole origin and this generally leads to the most rapid convergence of the multipole expansion.

Chapter 3

Theories and Models of the Liquid State

3.1 INTRODUCTION

A successful theory of the liquid state will be one that explains the bulk properties of the liquid from a knowledge of the intermolecular potential. In the bulk we are dealing with a very large number of molecules ($\sim 10^{25}\,\text{l}^{-1}$) and we know from the principles of statistics that at equilibrium the time fluctuations of bulk properties will be sufficiently small to be generally ignored. Statistical thermodynamics provides the analytical connection between the intermolecular potential and thermodynamic observables such as pressure and free energy. Non-equilibrium statistical mechanics provides a similar connection for time-dependent properties such as viscosity and thermal conductivity.

The statistical equations can be solved exactly or with high accuracy for gases and solids but, for reasons which will be explained later, they are very difficult to solve for liquids. Indeed, a rather advanced knowledge of statistical mechanics is needed to understand the most successful statistical theories of liquids and we shall make only passing comments on these in this chapter. Even the statistical theory of the imperfect gas, which we shall cover in detail, requires a good grounding in statistical thermodynamics and readers without this are advised to omit Sections 3.3 and 3.4.

An early alternative approach to a theory of liquids was through the use of physical models and in the next section we will describe work along this line that was undertaken by Bernal and others. These models have improved our conceptual understanding of the nature of liquids but they have not proved very helpful for the quantitative prediction of liquid-state properties.

In recent years a method for modelling liquids has been developed which is conceptually valuable and capable of giving a quantitative interpretation of bulk properties. This method uses a modern computer with fast operation and large core store to simulate the behaviour of a large number of atoms or molecules interacting under some assumed potential function. No computer is large enough (nor likely to be) to handle directly the interaction of 10^{23} particles but, as we shall see, techniques exist for dealing with much smaller numbers (10^2–10^3) which will give results not significantly different from the true bulk. These techniques will be described in Section 3.5.

Before the era of computer simulation there was considerable interest in relatively simple statistical models of the liquid state which were extensions of the statistical theories of solids. The *cell theory* of Lennard-Jones and Eyring's *significant liquid structure theory* are two of these. It cannot be said that they have an important place in modern liquid state theory because it is now accepted that the imperfect solid is a poor approximation to a liquid. The interested reader is directed to other books for an account of these theories.†

The treatment of liquids as imperfect gases is also not without serious criticism, but it at least has the merit of being relevant to fluids above the critical state. For this reason we give more attention to the statistical theory of the imperfect gas in Section 3.3 than might be thought justified in a book on liquids. We must emphasize, however, that normal liquids are neither imperfect gases nor imperfect solids. There is at present no theory of the liquid state which is satisfactory for all types of liquid and all conditions of temperature and pressure.

3.2 PHYSICAL MODELS

The physical models developed by Bernal and others are based on the proposition that the structure of liquids can be represented by the random close packing of hard or almost hard shapes. If one is thinking of a monatomic fluid, for example, then the model might be a cluster of ball bearings which has not been shaken down to a regular close-packed structure.

One of the first physical models examined by Bernal was a cluster of plasticine balls.‡ The relationship between the balls was determined by compressing the cluster so that the balls developed planar contact faces, that is, they became irregular polyhedra. When this model was dismantled it was found that the average number of faces of these polyhedra was 13.6 and the most commonly occurring face was pentagonal. A regular close-packed arrangement of spheres would give 12-faced polyhedra with rhombic faces. Only tetrahedra, cubes and octahedra of the regular solids can be packed in

---

†H. Eyring and M. K. Jhon, *Significant Liquid Structures*, Wiley, London and New York (1969); J. A. Barker, *Lattice Theories of the Liquid State*, Pergamon Press, Oxford (1963).
‡The three dimensional random packing of gelatine spheres was earlier examined by W. E. Morrell and J. H. Hildebrand, *J. Chem. Phys.*, **4**, 224 (1936), and before that others had studied packing in two dimensions. Bernal, however, was the first to emphasize the geometrical features of random close packing.

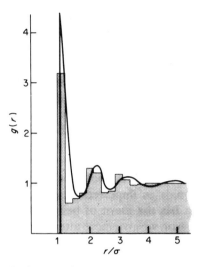

Figure 3.1. The radial distribution function for an assembly of hard spheres of diameter σ at high density. The continuous curve is the result of a computer simulation such as will be described in Section 3.5 and the histogram is the result of a ball-bearing experiment by Bernal and King. The figure was taken from J. S. Rowlinson, *Disc. Faraday Soc.*, **49**, 30 (1970)

an ordered array to fill space completely; polyhedra with pentagonal faces such as the dodecahedron cannot. Bernal therefore inferred that random close packing and pentagonal arrangements were necessarily connected.

Further examination of a randomly close-packed steel ball model, by Bernal and by Scott, led to a radial distribution function for hard spheres which is in good agreement with that found by computer simulation as shown in Figure 3.1. The coordination numbers in the model, defined by the number of neighbours whose centres were less than 1.05 diameters from a chosen atom, were found to vary between four and eleven with an average of nine. This agrees well with the results of computer simulation and the experimental value for neon determined by diffraction studies which is 8.8.

Scott's steel ball experiments led to a density for random close packing of 0.86 times the density of regular close packing and this also agrees well with, for example, the value of 0.87 for the ratio of the densities of solid and liquid argon at its melting point.

3.3 STATISTICAL THERMODYNAMICS

Statistical thermodynamics provides a scheme by which thermodynamic observables can be calculated as averages over molecular properties. In this chapter we first show how it may be applied to calculate the properties of the imperfect gas as this establishes the route for determining empirical potentials as described in the previous chapter. The analysis of the imperfect gas also shows one way that progress can be made for liquids.

All thermodynamic functions can be obtained from a quantity called the partition function which plays a central role in statistical mechanics. If a system has a set of discrete energy levels E_i, the partition function is defined by

$$Q = \sum_i \exp(-E_i/kT). \tag{3.1}$$

Q is a measure of the distribution of systems over the energy levels E_i. If these energy levels are very widely spaced compared with kT, only the lowest level, the energy of which can be taken as zero ($E_1 = 0$), will contribute to the sum in (3.1), and $Q = 1$. Very roughly, the value of Q is the number of states that are significantly populated at the temperature specified.

For an ensemble of systems with the same volume which are in thermal equilibrium the probability that a specified member has an energy E_i is†

$$P(E_i) = \exp(-E_i/kT)/Q. \tag{3.2}$$

Discrete energy levels arise in quantum mechanics for the electronic, vibrational and rotational states of molecules and for the translational states of atoms and molecules confined to finite volumes. By taking the system to be an individual molecule and substituting the appropriate expressions for these energies in to equation 3.1 one can deduce the molecular partition function and the thermodynamic stability of the species through the expression

$$A - A(T = 0) = -kT \ln Q, \tag{3.3}$$

where A is the Helmholtz free energy.

This procedure takes no account of the intermolecular potential. As the total energy for N molecules is not the sum of N separate molecular energies it is not strictly justified to take the system (in the statistical mechanical sense) to be an individual molecule, rather, it must be a set of N interacting molecules. In this case to make progress in the evaluation of equation 3.1 it is necessary to write the total energy as a sum of the internal energies, $E_{k,A}$, of the molecules (electronic, vibrational and rotational), the translational energies of the molecules and U the intermolecular potential

$$E_i = \sum_A (E_{k,A} + E_{trans,A}) + U. \tag{3.4}$$

---

†Terms such as 'system' and 'ensemble' have precise meanings in statistical mechanics. We direct the reader to specialist books on the subject for these and also for the derivation of 3.2 and the expressions relating Q to thermodynamic observables. The following two books are recommended: F. C. Andrew, *Equilibrium Statistical Mechanics*, Wiley, London and New York (1963), G. S. Rushbrooke, *Introduction to Statistical Mechanics*, Oxford (1949).

Expression 3.4 implicitly contains an important assumption which is that the intermolecular potential U is independent of the internal state (labelled k) of the molecules. For most molecules at normal temperatures only the lowest (ground) electronic state contributes significantly to the partition function and U is not very dependent on the vibrational or rotational states of the molecules. Without serious error U can be taken as the intermolecular potential appropriate to all molecules in their ground states.

The Configurational Integral

Because we have, by expression 3.4, written the total energy as a sum of component energies, the partition function 3.1 for a system of N molecules can be expressed as a product of component partition functions.† It can be shown that Q has the form

$$Q = \left(\frac{q_{int}q_t}{V}\right)^N \frac{Q'}{N!} \tag{3.5}$$

where q_{int} and q_t are the internal and translational partition functions for a single molecule and Q' is called the configurational integral, identified in expression 3.7. The internal partition function depends on the vibrational and rotational energies of the molecule, and it can be evaluated readily using the harmonic-oscillator and rigid-rotor models. However, as it has no role in our later analysis we have no need to quote its explicit form.

The translational partition function can easily be evaluated and for molecules of mass m contained in a volume V the result is

$$q_t = (2\pi mkT/h^2)^{3/2}V. \tag{3.6}$$

We note therefore that q_t/V is independent of volume and hence the volume dependence of Q is the same as the volume dependence of Q'.

A final point to note about expression 3.5 is the appearance of $N!$ in the denominator. This is associated with the indistinguishability of identical molecules in the system. However, the factor is only appropriate for identical molecules under conditions in which they can rapidly interchange their positions (i.e. in gases and liquids) and it must be omitted if this is not the case (i.e. for solids). In principle, identical molecules which have fixed places in a lattice are distinguishable.

The configurational integral is defined by

$$Q' = \int \cdots_{3N} \int \exp\left(-U/kT\right) dq, \tag{3.7}$$

---

†This is because of the mathematical identity

$$\exp\left(-\sum_i a_i\right) = \prod_i \exp\left(-a_i\right).$$

where U is the total intermolecular potential energy appearing in equation 3.4. The integration is over the $3N$ coordinates that define the positions of the centres of mass of the molecule and any rotational coordinates of the molecules which are parameters in the intermolecular potential.†

If N is the number of molecules in a macroscopic sample (say 10^{23}) there will be no hope of evaluating this integral unless it can be factored into products of integrals of smaller dimension. At first sight it might appear that this is relatively straightforward because, as argued in Section 2.1, the intermolecular potential is predominantly composed of pair potentials with smaller contributions from three-body and higher terms. Unfortunately this does not, of itself, lead to factorization of the configurational integral, as can be seen from the following example.

Consider a system of three atoms interacting with pair potentials $U(R_{ij})$ so that the total interatomic potential is

$$U = U(R_{12}) + U(R_{23}) + U(R_{13}). \tag{3.8}$$

The variables of integration are the coordinates of the centres of mass of the atoms whereas the R_{ij} are the distances between the centres, the two being related by expressions such as

$$R_{12} = ((x_2 - x_1)^2 + (y_2 - y_1)^2 + (z_2 - z_1)^2)^{1/2}. \tag{3.9}$$

Thus if we integrate over the position coordinates of atom 1, the part of the integral which depends on the position of atom 1 is

$$\exp((-U(R_{12}) - U(R_{13}))/kT), \tag{3.10}$$

and the resulting integral will be a function of the positions of atoms 2 and 3. For a system of N atoms, integration over the position coordinates of any one atom does not lead to a constant but to a function which depends on the positions of the other $N - 1$ atoms; in summary, the total integral cannot be expressed as a product of independent integrals.

In contrast, we note that for solids the situation is much simpler, because it can normally be assumed that U is a function (usually a harmonic function) of the displacement of an atom from its lattice site, and it does not depend explicitly on the distances between the atoms. Thus one can write

$$U = \sum_i U(q_i), \tag{3.11}$$

whence 3.7 reduces to a product of N single-atom integrals

---

†A qualification must be made that 3.7 is only valid when the motion of the centres of mass of the molecules can be treated by classical mechanics. The existence of negative regions (van der Waals' minima) of the intermolecular potential as illustrated in Figure 2.8 will be associated with discrete bound states. Except for very light atoms or molecules these discrete states are sufficiently close together that they can be treated as a continuum and classical mechanics applied. For helium the corrections for quantum behaviour are very important; for light atoms such as neon they are small and standard procedures exist for making a quantum correction.

$$Q' = \prod_{i=1,N} \int \exp\left(-U(q_i)/kT\right) dq_i, \tag{3.12}$$

which can be evaluated knowing the functional form of $U(q_i)$. An extension of this result was developed into the cell theory of liquids by Lennard-Jones and co-workers.

The Cluster Expansion

The way forward for liquids and gases is to make an expansion of the integrand of the configurational integral. To illustrate this approach we again assume that the total potential is pair-additive and write

$$U = \sum_{ij(j<i)} U_{ij}, \tag{3.13}$$

where $U_{ij} \equiv U(R_{ij})$ is a pair potential. A function f_{ij} called the Mayer function is defined by

$$f_{ij} = \exp\left(-U_{ij}/kT\right) - 1, \tag{3.14}$$

whence, from 3.7 and 3.13,

$$Q' = \int \cdots_{3N} \int \prod_{ij} (1 + f_{ij}) \, dq. \tag{3.15}$$

The product in 3.15 can now be expanded to give

$$Q' = \int \cdots_{3N} \int \left(1 + \sum_{ij} f_{ij} + \sum_{ij} \sum_{kl \neq ij} f_{ij} f_{kl} + \ldots \right) dq, \tag{3.16}$$

which is called the cluster expansion. The first term is independent of the intermolecular potential or, giving it another perspective, is the only term which appears in the ideal gas limit when the intermolecular potential is zero. The terms linear in f represent the interaction of molecules in pairs, and the higher terms represent the interaction of molecules three or more at a time.

The first term in the cluster expansion is easily evaluated. For interacting atoms the $3N$ dimensional integral over the complete volume V which they occupy is equal to V^N. The second term is evaluated by first noting that all the U_{ij} (hence all the f_{ij}) are equivalent; therefore each of the $N(N-1)/2$ terms of this kind must give the same contribution to the integral. Taking f_{12} as a typical term and integrating over the coordinates of all other atoms leads to the following expression for Q'

$$Q' = V^N + \tfrac{1}{2} N^2 V^{N-2} \int \int f_{12} \, dq_1 \, dq_2 + \text{higher terms}, \tag{3.17}$$

where we have replaced $N(N-1)$ by N^2, a justified approximation for large N. The higher terms are evaluated in a similar way and reduce to integrals over the positions of three or more atoms.

The b Integral

For atomic systems U_{ij} is a function only of R, the distance between the atoms, and no orientational variables are involved. The integral over f_{12} in 3.17 can in this case be reduced further (see Appendix 3.1) to give the following result

$$Q' = V^N\left[1 - \frac{2\pi N^2 b}{V} + \ldots\right], \tag{3.18}$$

where

$$b = -\int_0^\infty f_{12}(R)R^2\,\mathrm{d}R = \int_0^\infty (1 - e^{-U_{12}(R)/kT})R^2\,\mathrm{d}R. \tag{3.19}$$

The integral b can readily be evaluated by numerical techniques if the interatomic potential $U(R)$ is known, and it has been extensively studied for the L-J potential (2.18). One of the few potentials for which an analytical integration is possible is the hard-sphere potential (Figure 2.8(a)). For this potential, $U(R)$ is infinite in the region $0 < R < \sigma$ and integration in this range gives $\sigma^3/3$. For the region $R > \sigma$, $U(R)$ is zero and so in this range there is no contribution to the integral. Thus for the hard-sphere model we have $b = \sigma^3/3$ and

$$Q' = V^N\left[1 - \frac{2\pi N^2 \sigma^3}{3V} + \ldots\right]. \tag{3.20}$$

Because σ^3 is of the order of a molecular volume, (V/N), the second term in the brackets of (3.20) must be of the order of N, $\sim 10^{23}$ for a macroscopic sample. Moreover, higher terms in the cluster expansion 3.16 are of the magnitude of higher powers of N, so that the cluster expansion as it appears in Q' is extremely divergent and there would be no justification for terminating it at the second term. However, thermodynamic observables depend on $\ln Q$, as can be seen from 3.3. The contribution from Q' to the Helmholtz free energy is, from 3.18

$$A - A(T = 0) = -kT\ln Q' = -kTN\left[\ln V + N^{-1}\ln\left(1 - \frac{2\pi N^2 b}{V} + \ldots\right)\right], \tag{3.21}$$

and because a logarithmic function is a very slowly divergent function of its argument, the cluster expansion for A can be shown to be convergent, at least for dilute gases.† In fact the correct result can be obtained by treating

---

†The full analysis is quite complicated and was first given by Ursell, *Proc. Camb. Phil. Soc.*, **23**, 685 (1927).

the second term in the cluster expansion as small and using the approximation $\ln(1 + x) = x$ for small x. In this limit 3.21 becomes, for one mole (with $N = \mathcal{N}$, $k\mathcal{N} = R$)

$$A - A(T = 0) = -RT\left(\ln V - \frac{2\pi\mathcal{N}b}{V_m}\right). \tag{3.22}$$

Of particular interest is the pressure of the system which can be obtained from the relationship

$$P = -\left(\frac{\partial A}{\partial V}\right)_T. \tag{3.23}$$

Differentiating 3.22 we obtain the result

$$P = \frac{RT}{V_m}\left(1 + \frac{2\pi\mathcal{N}b}{V_m}\right). \tag{3.24}$$

If b is zero, 3.24 is the perfect gas equation. The term in b is a correction which approaches zero as the molar volume increases. Analysis of further terms in the cluster expansion and their contribution to P shows that 3.24 comprises the first two terms in an expansion for P in powers of $1/V_m$. These terms will be related to an empirical equation for the pressure in Section 3.6.

We can now return to an examination of the size of the b-term and its contribution to the pressure using the hard-sphere model for simplicity ($b = \sigma^3/3$). Taking argon as our example, σ is approximately 3.4 Å. The molar volume of gaseous argon at 273 K and 1 atm is 22.4 dm^3 and the molar volume of liquid argon at 87 K is 28.6 cm^3. From these data one obtains values for the dimensionless quantity $2\pi\mathcal{N}b/V_m$ of 2.2×10^{-3} for the gas, and 1.7 for the liquid. The b-term is therefore a small correction to the pressure for the gas, but a large correction for the liquid and the cluster expansion terminated at the first correction term is unlikely to give the correct pressure for the liquid state. We will return to an examination of expression 3.24 for gases in Section 3.6 in order to show how experimental measurement of (P, V, T)-properties can be used to deduce b, and hence allow parameterization of the intermolecular potential.

Although it is clear from the above analysis that the cluster expansion 3.16 terminated at the linear terms is inadequate for liquids, the ideas which are behind it have proved valuable. J. E. Mayer and others have made an extensive analysis of the cluster integrals using graph theory techniques to reduce and sum the terms.† Although these techniques have not been sufficient to provide a direct solution to the configurational partition function (3.16) for liquids they have proved useful in alternative statistical approaches to liquids based on distribution functions, and it is this aspect that we explore in the next section.

---

†J. E. Mayer and M. Goeppert-Mayer, *Statistical Mechanics*, Wiley, New York (1940).

3.4 DISTRIBUTION FUNCTIONS

In discussing the concept of order in a liquid in Section 1.3 we introduced an important function in the theory of liquids called the radial distribution function, $g(r)$. This and similar functions play a central role in the more advanced statistical theories of liquids and, although we will not be describing such theories in great detail, it is important to see the general direction in which the subject has progressed.

The instantaneous state of a system of N particles in classical mechanics is described by specifying the $3N$ momenta and $3N$ position variables of the particles. We start by defining a function $f^{(N)}$ which gives the probability of finding the particles in a specified classical state. The probability that particle 1 will have momentum between p_1 and $p_1 + dp_1$ and position between q_1 and $q_1 + dq_1$, and simultaneously particle 2 has momentum between p_2 and $p_2 + dp_2$, etc., is

$$f^{(N)}(p_1 \ldots p_N, q_1 \ldots q_N) \, dp_1 \ldots dp_N \, dq_1 \ldots dq_N. \qquad (3.25)$$

To obtain a general N-particle configurational probability function which describes the positions irrespective of momenta, we can integrate over the momentum variables to obtain a quantity $n^{(N)}$ that depends only on the coordinates of the particles

$$n^{(N)}(q_1 \ldots q_N) = \int \cdots_{3N} \int f^{(N)}(p_i, q_i) \, dp_1 \ldots dp_N. \qquad (3.26)$$

However, even this function has no direct relevance to experiment when N is large, because no experiment can give information about the simultaneous positions of all particles in the system. What can be deduced is some information about the relative positions of small numbers of particles and particularly, as we have seen in Section 1.3, about the distribution of pairs of particles.

The distribution functions for small numbers of particles can be obtained from 3.26 by successive integrations over the position variables. Thus if we know $n^{(N)}$ we can find by integration $n^{(N-1)}$, $n^{(N-2)}$, etc., down to $n^{(1)}$. The general recurrence formula is

$$\int n^{(k+1)} \, dq_{k+1} = (N - k) n^{(k)}, \qquad (3.27)$$

which means that the integral is equal to the distribution function for K particles multiplied by the number of particles encountered in performing the integration over the whole volume. In an isotropic fluid the probability of finding a particle in a volume dq_1 is independent of q_1, hence we have

$$n^{(1)} = N/V \text{ and } \int n^{(1)} \, dq_1 = N. \qquad (3.28)$$

From 3.27 the integral of the pair distribution function is given by

$$\int n^{(2)}(\boldsymbol{q}_1, \boldsymbol{q}_2) \, \mathrm{d}\boldsymbol{q}_2 = (N - 1)n^{(1)} = N(N - 1)/V. \tag{3.29}$$

The radial distribution function $g(r)$ is related to $n^{(2)}$ by the expression

$$n^{(2)}(\boldsymbol{q}_1, \boldsymbol{q}_2) = n^{(1)}(\boldsymbol{q}_1)n^{(1)}(\boldsymbol{q}_2)g(r) = (N^2/V^2)g(r). \tag{3.30}$$

As r approaches infinity, $g(r)$ approaches unity, and in this limit the probability of finding one particle at $\boldsymbol{q}_1$ and a second at $\boldsymbol{q}_2$ is the product of the two one-particle probabilities.

For a system in equilibrium $n^{(N)}$ will be proportional to $\exp(-U/kT)$, where U is the total potential energy, as defined by 3.4, for the N particles. The proportionality constant is equal to $N!/Q'$, where Q' is the configurational integral (equation 3.7). This can be confirmed by integrating both sides of the expression

$$n^{(N)} = \frac{N!}{Q'} \exp(-U/kT), \tag{3.31}$$

over the coordinates of all N particles and using the integration formula 3.27. By successive integrations over the coordinates of each particle, expressions for the lower order distribution functions can be obtained and, in particular,

$$n^{(2)} = \frac{N!}{(N-2)!Q'} \int \ldots \int \exp(-U/kT) \, \mathrm{d}\boldsymbol{q}_3 \ldots \mathrm{d}\boldsymbol{q}_N. \tag{3.32}$$

Thus to calculate these distribution functions from the intermolecular potential requires a similar, and as difficult, an integration as does the calculation of Q'.

Integral Equations

The advanced statistical theories to which we have referred earlier provide approximations to the distribution function $n^{(2)}$ and to $g(r)$. One of the most important ideas that has been used is contained in the Kirkwood super-position approximation

$$n^{(3)}(\boldsymbol{q}_1, \boldsymbol{q}_2, \boldsymbol{q}_3) = \frac{V^3}{N^3}n^{(2)}(\boldsymbol{q}_1, \boldsymbol{q}_2)n^{(2)}(\boldsymbol{q}_2, \boldsymbol{q}_3)n^{(2)}(\boldsymbol{q}_3, \boldsymbol{q}_1). \tag{3.33}$$

This states that the probability of finding particles at $\boldsymbol{q}_1$, $\boldsymbol{q}_2$, $\boldsymbol{q}_3$, is proportional to the product of the three pair probabilities. The equation is exact if the three particles are far apart, and from this limit one deduces that the factor $(V/N)^3$ must be included for correct dimensionality. The equation will not be exact if the particles are close together (if 1 and 2 are close together this will influence the probability that 3 is close to 1), but the errors arising from the approximation do not appear to be large. Integration of both sides of 3.33 over the position variables of one particle leads on the left to $n^{(2)}$ (by 3.27), and on the right to an integral involving $n^{(2)}$. This is therefore an integral equation for $n^{(2)}$ and such equations have been developed and analysed by Born, Green and Yvon.

The other important direction that has been followed in the statistical mechanics of fluids starts from an expansion of $\ln g(r)$ in powers of $1/V$, using similar techniques for analysing clusters of particles to those of Mayer which were described in the last section. Two theories, known as the hyper-netted chain (HNC) and Percus-Yevick (PY) theories have been developed in this way and are still in active use at the research level. For the derivation and analysis of these equations we refer the reader to more advanced texts.†

Finally we note that for molecules there is interest not only in the distribution functions for the centres of mass, but also in distribution functions for orientation. For example, for an assembly of linear molecules one would be interested in the probability that a second molecule has its centre of mass at a distance R from the first, and that it subtends angles of θ_a, θ_b, ϕ, as defined in Figure 2.4. Distribution functions containing this information are particularly important for analysing the liquid–crystal phase which will be described in Chapter 5.

Velocity Correlation

We saw in Section 1.3 that gases, liquids and solids are distinguished by having very different radial distribution functions. Spatial distribution functions describe the time-averaged relative positions of molecules in the phase. The way that the position or velocity of a molecule evolves with time is also characteristic of the phase. In gases molecules move rapidly through large distances and are deflected only on collision with other molecules. In solids the molecules oscillate about their lattice sites and only rarely jump between sites. In liquids the molecules move around like dancers in a crowded ballroom; they progress slowly with frequent collisions.

The function which brings out most clearly differences in the time evolution of gases, liquids and solids is called the velocity autocorrelation function. It is defined by

$$\psi(t) = \langle \boldsymbol{v}(0) \cdot \boldsymbol{v}(t) \rangle / \langle \boldsymbol{v}(0)^2 \rangle, \tag{3.34}$$

where $\boldsymbol{v}(0)$ is the velocity of a molecule at a particular time and $\boldsymbol{v}(t)$ the velocity of the same molecule at a time t later. The brackets indicate averages over all molecules in the system; $\langle \boldsymbol{v}(0)^2 \rangle$ is the average squared velocity of the molecules.

At $t = 0$, $\psi(t)$ is equal to unity. When t becomes very large the molecule may still be moving with a speed similar to that at $t = 0$, but the direction in

---

†C. A. Croxton, *Introduction to Liquid State Physics*, Wiley, Chichester (1975). J. P. Hanson and I. R. McDonald, *Theory of Simple Liquids*, Academic Press, London and New York (1976).

52

which the molecule is travelling will be completely unrelated to its direction at $t = 0$ because of all the collisions that have occurred in this time. Because the numerator of 3.34 is a scalar product (i.e., proportional to the cosine of the angle between the vectors $\boldsymbol{v}(0)$ and $\boldsymbol{v}(t)$) the average value of this scalar product will approach zero as t approaches infinity. However, the rate and manner in which it approaches zero is very different for gases, liquids and solids.

In a dilute gas the mean free path of a molecule (the average distance between collisions) is large and on collision a molecule is more likely to suffer a small deflection than a large deflection. For this reason $\psi(t)$ approaches zero slowly and smoothly. In a solid a molecule will oscillate about its lattice site and for a simple harmonic motion $\psi(t)$ will also have harmonic form with the same frequency as the vibration.

In a liquid the velocity autocorrelation function starts initially like that of a solid, showing an oscillatory form due to the collision of a molecule with its neighbours. However, the motion is not harmonic and the molecule has a short memory of its initial velocity. Thus $\psi(t)$ is oscillatory but the amplitude of the oscillations decays quite rapidly to zero.

Figure 3.2 shows the typical behaviour of $\psi(t)$ that we have described above. The liquid state function is one that has been calculated for liquid CO using the method of computer simulation to be described in the next section.

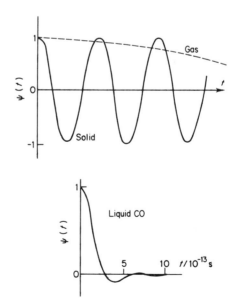

Figure 3.2. The velocity autocorrelation function for typical solid, liquid and gas. The liquid curve is a calculation by B. J. Berne and G. D. Harp, *Adv. Chem. Phys.*, **17**, 63 (1970)

3.5 COMPUTER SIMULATION†

With the present generation of large electronic computers it is possible to solve the classical equations of motion for up to about 10^4 particles interacting through some analytical pair potential of the type described in Section 2.4, or for fewer particles with more complicated potentials. One can then calculate the time-average or time-dependent properties of this system and compare these with the properties of a real fluid. If it can be established that the number of particles considered is large enough to represent a macroscopic sample, then the comparison between computer experiment and physical experiment provides a measure of the accuracy of the pair potential and of the assumption that the total potential is pair additive.

Of course 10^4 particles will not behave in the same way as 10^{23} particles. Such a small sample would have a large surface area relative to its volume and, as we shall see in Chapter 13, molecules at the surface have different energies from molecules in the bulk. We can, however, get around this difficulty by supposing that our particles are confined in a small cubic volume and that this volume, together with its pattern of positions and momenta of the particles, is translated with infinite repetition along the three edges of the cube. In other words the elementary cube has the same relationship to the whole fluid as does a unit cell of a crystal to the whole crystal lattice. An instantaneous picture of a two-dimensional fluid in this model is shown in Figure 3.3. The arrows associated with each point indicate the momenta of the particles.

It is clear from Figure 3.3 that we have not modelled a true fluid by our translational operation because, as has been stressed in Section 1.3, a fluid has no long-range order. However, providing the range over which the interparticle potential acts is less than half the side length of the cell (d) then the total potential in which a particle finds itself is not influenced by the artificial translational symmetry. Thus particle A_5 might interact with B_5 but not with B_4 or B_6 which are more than $d/2$ away. Similarly A_5 can interact with C_2 but not with C_5 etc. In other words particle A is influenced by the potential of only one B, C, D . . . particle.

For short-range potentials, we can therefore simulate a fluid by a relatively small number of particles at normal densities and indeed the first successful computer calculations by Alder and Wainwright in 1959 on hard spheres used only 32 particles in the cell. For long-range potentials it is necessary to have large cells and hence at normal densities large numbers of particles. In particular, for ions there is the difficulty that the Coulomb potential between equivalent particles diverges, and there are problems with convergence even for dipolar molecules. Techniques have been devised to

---

†M. P. Allen and D. J. Tildesley, *Computer Simulation of Liquids*, Clarendon Press, Oxford, (1987).

54

account for these long-range interactions, but we will not elaborate the point.

A further point that can be seen from Figure 3.3 is that during the trajectory a particle will encounter the cell wall; particle B is an example of this. However, if the cell walls have no physical reality then as one particle passes out of a cell (e.g., B_5 from the central cell in Figure 3.3) then another (e.g., B_6) will enter from the other side. Thus the model conserves particle density when this is averaged over the elementary cell.

Molecular Dynamics

The amount of computer time that is required to simulate the dynamics of 10^4 molecules can be large even on the fastest computers. Small molecules in liquids randomize their positions and orientations in about 10^{-12} s, so that one needs to simulate about 10^{-10} s of physical time to obtain an average representation of the liquid, and this may require several hours of computing time. For molecules which have complicated potentials this time would have to be increased considerably, or the number of molecules reduced, for calculations to be feasible. Non-rigid molecules (i.e. those which are allowed to vibrate) present another order of magnitude of difficulty.

Molecular dynamics calculations are carried out using standard numerical integration procedures for coupled differential equations. An initial starting configuration is chosen, usually a regular array, and momenta assigned in a

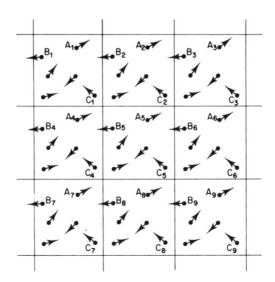

Figure 3.3. Construction of a macroscopic fluid from an elementary cell (two dimensions). Note that particle labelled A, for example, has the same relative position and momentum in each cell

random manner subject to a restraint provided by the total energy of the system. After a relatively short computing time the transient effects of the starting configuration will have died out and this part of the trajectory is discarded as having no physical significance. From then on the positions and momenta of the particles are recorded as a function of time and used to calculate time-dependent or time-averaged properties, in particular the radial distribution function.

Phase transitions are not easy to reproduce by computer simulation because long-range correlation is artificially built into the model as we have noted, yet one should record for historical interest the important result obtained by Alder and Wainwright that both solid and fluid phases can be reproduced for hard spheres and that the phase change occurs when the volume V is approximately 1.4 times the close packed volume V_0.[†]

Figure 3.4 shows the typical trajectories obtained in the solid and liquid phases and Figure 3.5 shows the pressure–volume relationships deduced in the high and low density regions. If V/V_0 is close to 1.4 then starting with a regular configuration the trajectories are solid-like, but if one starts with an irregular configuration they are liquid-like.

Monte Carlo Method

To calculate equilibrium configurational properties of a fluid it is not, of course, necessary to know the time-dependence of the particle motion; it is sufficient to have just a selection of configurations from which an appropriate average can be extracted. The fact that a trajectory passes through a set of configurations allows this method to be used for configurational averages, although inherent in this is the so-called ergodic assumption that time averages of dynamical systems are equivalent to statistical mechanical averages. A less costly method of calculating these averages on a computer is to construct a random set of configurations and average over them without bothering about the momentum variables or the time evolution of a configuration.

It is possible to construct a random configuration for N particles in a cubic box with sides $2k$, by selecting $3N$ random numbers and to use these to define the positional coordinates of the particles. For example if the first random number is ζ, lying between the limits 0 and 1, then an x coordinate for particle 1 given by

$$x = (2\zeta - 1)k, \tag{3.35}$$

will lie somewhere in the limits $-k \leqslant x \leqslant k$; 2nd and 3rd random numbers give its y and z coordinates. Such an approach to the construction of a random set of configurations is referred to as the Monte Carlo method, the connotation with games of chance being evident.

[†]B. J. Alder and T. E. Wainwright, *J. Chem. Phys.*, **31**, 459 (1959).

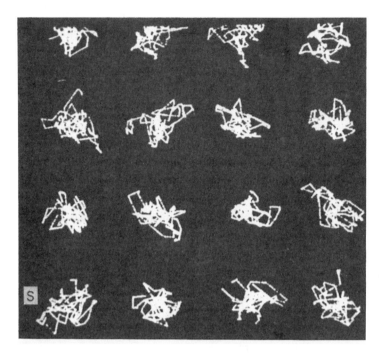

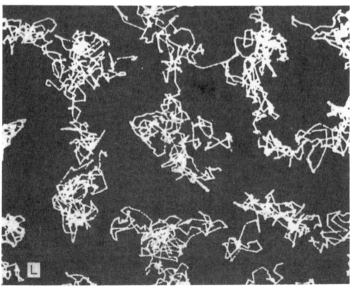

Figure 3.4. Trajectories of hard spheres in the solid (S) and liquid (L) phases from the results of Alder and Wainwright. Calculations were on 32 particles which underwent approximately 3000 collisions. Only the trajectories of half the particles are shown. We are indebted to Professor B. J. Alder for providing original photographs for this figure

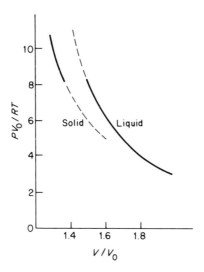

Figure 3.5. Pressure–volume relationship for hard spheres as deduced by molecular dynamics calculations. The solid–liquid phase change occurs at approximately $\dot{V}/V_o = 1.4$

The average value of a property $X(q_i)$, which is a function of the coordinates of the particles, in a system that obeys Boltzmann statistics, is

$$\bar{X} = \frac{\int e^{-U(q_i)/kT} X(q_i) \, dq_i}{\int e^{-U(q_i)/kT} \, dq_i}, \tag{3.36}$$

where $U(q_i)$ is the potential energy. If we therefore generate a large number, M, of configurations and calculate the value of X for each, $\bar{X}$ can be approximated by

$$\bar{X} = \sum_{i=1}^{M} P_i X_i, \tag{3.37}$$

where

$$P_i = \frac{e^{-U_i/kT}}{\sum_{i=1}^{M} e^{-U_i/kT}}, \tag{3.38}$$

is the probability of configuration i arising. Calculating $\bar{X}$ in this way is called the crude Monte Carlo method.

The crude Monte Carlo method described above for choosing and weighting the configurations is extremely inefficient because most configurations produced are highly improbable for a real fluid. Most configurations produced by a set of random numbers will have some atoms lying on top of one another, that is, some particles will be so close together that the potential energy U_i will be extremely high. What is needed therefore is a method of selecting configurations directly so that they occur in the set with probability P_i, and if this is the case then we can replace 3.37 by

$$\bar{X} = \frac{1}{M}\sum_{i=1}^{M} X_i. \tag{3.39}$$

A procedure for achieving this was proposed by Metropolis and co-workers in 1953, and it is this which, for liquid state computer simulation, is generally called the Monte Carlo method. A detailed description of the procedure is given in Appendix 3.2.

Apart from the confirmation of important concepts of the fluid state by computer simulation, it should be emphasized that many measurable quantities can be reproduced accurately given a suitable pair potential. Thus even early calculations by Wood and Parker in 1957 using small numbers of particles showed that, with a Lennard-Jones potential for argon, with parameters fitted to the second virial coefficient, the thermodynamic properties of liquid argon could quite accurately be reproduced. Some later results by McDonald and Singer are given in Table 3.1. The results can be improved by further refinement of the potential.

The computer simulation of molecular, as distinct from atomic, fluids is an active area of research, but there is good evidence to suggest that experimental data, particularly diffraction data, can be reproduced by relatively simple intermolecular potentials.† For example, a hard core model for CS_2 with parameters shown in Figure 3.6 has given good agreement with X-ray scattering data. We restate a point made earlier that for strongly polar

Table 3.1. P, V, T data for argon†

T/K	V_m/cm^3	P/atm exp.	P/atm calc.	$-\bar{U}/kJ\,mol^{-1}$ exp.	$-\bar{U}/kJ\,mol^{-1}$ calc.
97.0	28.48	214	200 ± 14	5.799	5.912
97.0	28.95	141	141 ± 10	5.736	5.807
108.0	28.48	451	433 ± 12	5.690	5.803
108.0	31.72	16	-16 ± 10	5.204	5.247
117.0	28.48	619	605 ± 16	5.581	5.740
117.0	29.68	386	399 ± 14	5.393	5.519
117.0	30.92	219	174 ± 13	5.209	5.347

†Calculated by I. R. McDonald and K. Singer, *J. Chem. Phys.*, **50**, 2308, (1969) from a Lennard-Jones pair potential. The statistical error in calculating the average configuration internal energy is estimated at $\pm 8 \times 10^{-3}\,kJ\,mol^{-1}$.

---

†Several papers on this topic were presented in 'Structure and motion in molecular liquids', *J.C.S. Faraday Disc.*, **66** (1978).

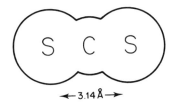

Figure 3.6. A hard-core model for CS_2. The diameters of the sulfur and carbon atoms are 3.5 Å and 3.0 Å, respectively. (S. I. Sandler and A. H. Narten, *Mol. Phys.*, **32**, 1543 (1976))

molecules the electrostatic contribution to the potential must certainly be allowed for and will probably have an important influence on the liquid structure.

3.6 EQUATIONS OF STATE

For a perfect gas the equation relating pressure, volume and temperature, is

$$P = RT/V_m. \tag{3.40}$$

For a real gas there are small deviations from this equation and for liquids it is not valid at all. However, in principle, more complicated equations can be found which represent the (P, V_m, T) behaviour through the whole fluid range and these can be written in the alternative forms

$$P = \psi(V_m, T) \quad \text{or} \quad \phi(P, V_m, T) = 0 \tag{3.41}$$

where ψ and ϕ are functions of the variables in brackets. These are called equations of state for the fluid.

Figure 3.7 shows the pressure–volume curves for CO_2 for three temperatures; one above, one below, and one at T_c, the critical temperature. These curves are typical of all fluids and, as we show later, by a suitable scaling of pressure, volume and temperature all fluids are found to fit quite closely to a common set of curves.

At temperatures above T_c there is only one fluid phase. At low pressures the P–V isotherm is hyperbolic, conforming to the perfect gas law (3.40), but at high pressures the fluid becomes much less compressible than a perfect gas and the isotherm rises much more steeply than does a hyperbola. At temperatures below T_c the isotherms are approximately hyperbolic (the deviation from the perfect gas law will later by examined in more detail) until a pressure is reached at which the first drop of liquid appears.

We have seen in Section 1.2 that when a single component has two phases in equilibrium there is only one degree of freedom, so that if this is used to specify the temperature, then the pressure is fully determined. In other words, no matter what amount of liquid we have in equilibrium with gas, the

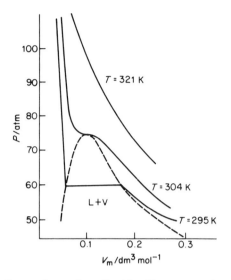

Figure 3.7. Three isotherms for carbon dioxide. The area under the broken curve is the region where liquid (L) and vapour (V) coexist

pressure is constant (at a given T) although the volume can vary. If, however, the volume is decreased until the system is totally liquid then the compressibility drops immediately to a very low value, that is, V changes very little with a further increase of pressure.

If a family of (P, V_m) isotherms is plotted at temperatures below the critical temperature, the points of discontinuity which mark the boundaries of the liquid–gas equilibrium fall on a curve which is shown by the broken line in Figure 3.7. The curve in its left-hand branch represents the molar volume of the liquid, and in its right-hand branch the molar volume of the gas when liquid and gas are in equilibrium. The top of the curve is the critical point at which these two volumes are equal, that is, the densities of the gas and liquid phases are equal at the critical point.

The (P, V_m) isotherm at the critical temperature is gas-like at low pressures and liquid-like at high pressures, and the critical point is a point of inflexion which satisfies the criteria

$$\left(\frac{\partial P}{\partial V}\right)_T = 0, \quad \left(\frac{\partial^2 P}{\partial V^2}\right)_T = 0, \quad \left(\frac{\partial^3 P}{\partial V^3}\right) < 0. \tag{3.42}$$

The first of these conditions means that $V^{-1}(dV/dP)_T$, the isothermal compressibility, is infinite at the critical point. In practice, the fluid is infinitely compressible only over an infinitesimal range of pressures.

The behaviour of other thermodynamic quantities as the critical point is reached is a matter of some complexity. The experimental result is that the Helmholtz free energy, A, is not an analytical function of V and T at the critical point, and the specific heat at constant volume, which is given by

$$C_V = (\partial U/\partial T)_V = -T(\partial^2 A/\partial T^2)_V \qquad (3.43)$$

is infinite. We refer the reader to more advanced texts for a more detailed discussion of the critical state.†

Van der Waals' Equation

The first quantitatively successful equation of state of a fluid that describes the (P, V_m) curves of the liquid and gaseous branches was that of van der Waals. This is usually written in the form

$$(P + a/V_m^2)(V_m - b) = RT, \qquad (3.44)$$

which can be rearranged to

$$P = \frac{RT}{V_m - b} - \frac{a}{V_m^2}. \qquad (3.45)$$

In this equation a and b are empirical parameters which can be given a physical interpretation. The parameter b, for example, can be taken to represent the nearly incompressible volume of randomly close-packed molecules. We see from 3.45 that this parameter occurs in the first term and has the effect of raising the pressure above its perfect-gas value. The parameter a is associated with the attractive forces between molecules and these will decrease the pressure below its perfect-gas value. The term a/V_m^2 is called the internal pressure of the fluid.

From these physical arguments one could expect some temperature dependence of both the van der Waals parameters. Experiment shows that temperature is more important for the attractive term, and an equation of state was proposed by Berthelot which contains a temperature-dependent internal pressure.

$$(P + a/TV_m^2)(V_m - b) = RT. \qquad (3.46)$$

This equation is considered to be a slight improvement over the van der Waals equation.

If the van der Waals equation is differentiated with respect to V and the critical point conditions 3.42 applied, one obtains the following relationships between the critical constants and the van der Waals parameters a and b:

$$P_c = a/27b^2, \quad T_c = 8a/27Rb, \quad V_{c,m} = 3b. \qquad (3.47)$$

Moreover, the ratio

$$Z_c = P_c V_{c,m}/RT_c, \qquad (3.48)$$

which is called the compression factor at the critical point, has the value

---

†For example, J. S. Rowlinson, *Liquids and Liquid Mixtures*, 2nd Edn, Ch. 3, Butterworth, London (1969).

Table 3.2. Values of the compression factor at the critical point and the van der Waals constants (*Handbook of Chemistry and Physics*, 59th Edn, CRC Press, 1978)

	Z_c	$a/\text{dm}^6\,\text{atm}\,\text{mol}^{-1}$	$100b/\text{dm}^3\,\text{mol}^{-1}$
He	0.306	0.034	2.370
Ne	0.308	0.211	1.709
H_2	0.305	0.244	2.661
N_2	0.291	1.390	3.913
CO_2	0.275	3.592	4.267
H_2O	0.227	5.464	3.049
NH_3	0.242	4.170	3.707
C_2H_6	0.285	5.489	6.380
C_6H_6	0.274	18.00	1.154

The van der Waals constants are not determined solely from critical point data according to 3.47, but are optimized for the whole (P, V_m, T) behaviour.

$3/8 = 0.375$, and hence is independent of a and b. Table 3.2 shows experimentally determined values of Z_c and the van der Waals parameters for a number of fluids. It is seen that Z_c is always a little less than the van der Waals value, the smallest values being for polar molecules.

Other equations of state can be analysed in a similar manner to the van der Waals equation. The Berthelot equation, for example, also gives $Z_c = 0.375$, but an equation due to Dieterici

$$P = \frac{RT}{(V_m - b)} \exp\left(-a/RTV_m\right), \tag{3.49}$$

leads to $Z_c = 0.271$, which is rather closer to most experimental values. However, Dieterici's equation is very poor at high pressures.

The equation of state for the hard-sphere model of a gas has been known quite accurately for many years. The Percus–Yevick equation, for example (see page 51), gives the equation of state

$$P = \frac{RT}{V_m}\left(\frac{1 + \eta + \eta^2}{(1 - \eta)^3}\right), \tag{3.50}$$

where $\eta = \pi\rho\sigma^3/6$; ρ is the density of the gas and σ is the hard-sphere diameter (Figure 2.8(a)). This equation agrees very well with pressures determined by computer simulation. Without attractive forces between the molecules there can be no liquid–gas phase change. Thus the hard-sphere model can only represent a fluid above its critical point. However, the hard-sphere model is a much more accurate representation of the effect on pressure of the short-range repulsive forces between molecules than is the first term in the van der Waals' equation (3.45). For this reason equations of state which are extensions of the hard-sphere result (3.50), can be very accurate. For example, Carnahan and Starling have proposed an equation

$$P = P_0 - \frac{a}{T^{1/2}V_m(V_m + b)}, \tag{3.51}$$

where P_0 is given by 3.50, which is very much better than any of the other two-parameter equations of state mentioned above.[†]

Law of Corresponding States

Although the critical constants of fluids cover very wide ranges (see Table 1.1), it is found empirically that the critical temperature is generally about 1.5 times the normal boiling point, and the critical volume is about three times the molar volume of the normal liquid. This suggests that if the pressure, volume and temperature are scaled to the values of the critical constants, then all fluids should obey, approximately, the same equation of state. This idea was first explored by van der Waals. He defined the reduced variables

$$P_r = P/P_c, \quad V_r = V_m/V_{c,m}, \quad T_r = T/T_c, \tag{3.52}$$

and noted that different gases at the same reduced volume and temperature exerted approximately the same reduced pressure. This is called the law of corresponding states.

We have already seen that the critical constants of fluids that obey the van der Waals equation are related to the parameters a and b by 3.47. It follows that one can replace a and b in the van der Waals equation by P_c, $V_{c,m}$ and T_c, and, after rearrangement, the equation can be written in reduced variables in the form

$$P_r = \frac{8T_r}{3V_r - 1} - \frac{3}{V_r^2}. \tag{3.53}$$

Figure 3.8 shows isotherms of the reduced van der Waals equations at three values of T_r. For $T_r > 1$ the isotherms have no turning points and have the general features of fluid isotherms above the critical point. For $T_r = 1$, the isotherm shows the required point of inflexion at $P_r = V_r = 1$. For $T_r < 1$, however, the isotherms have two turning points and they do not exhibit the constant pressure region which is associated with the gas–liquid equilibrium.

The van der Waals equation predicts that below the critical temperature an increase in pressure can lead to an increase in volume which is physically impossible. One can, however, obtain physically acceptable isotherms from the van der Waals equation by replacing the region having the two turning points by a straight line such that the two cut-off areas (shown in Figure 3.8)

[†]An excellent review of this and other equations of state is given by D. Henderson, Ch. 1, Equations of state in engineering and research, *Adv. in Chem.*, **182** (1979).

64

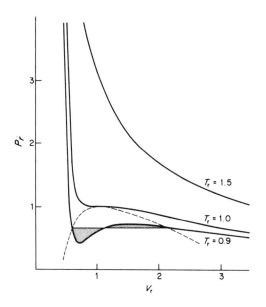

Figure 3.8. Isotherms of the reduced van der Waals equation

are equal.† This rule, known as Maxwell's equal area rule, is proved in Appendix 3.3.

Figure 3.9 shows the compression factor $Z = PV_m/RT$ at 273 K for some small molecules. At low pressures gases are more compressible than the perfect gas due to the long-range attractive forces between the molecules; at high pressures they are less compressible due to the short-range repulsive forces. If such data are plotted in reduced variables, they all fall close to the same curve. This is shown in Figure 3.10 for a large number of molecules and several different values of the reduced temperature.

One can attribute the law of corresponding states to the fact that similar molecules have similar functional forms for their intermolecular potentials. For example, suppose two molecules A and B have L-J potentials (2.18) but with different values for the parameters ε and σ.

$$V_{AA} = 4\varepsilon_A[(\sigma_A/R)^{12} - (\sigma_A/R)^6], \tag{3.54}$$

$$V_{BB} = 4\varepsilon_B[(\sigma_B/R)^{12} - (\sigma_B/R)^6]. \tag{3.55}$$

Because ε is an energy-scaling parameter in the Lennard-Jones potential and kT has dimensions of energy, we expect systems A and B to behave similarly with respect to their population of energy states if

---

†The van der Waals equation is analytical (i.e. differentiable to all orders) for all values of P, V and T, whereas the isotherms for real fluids are not analytical at the points of onset of the two-phase region. Thus the failure of the van der Waals equation to represent the isotherms in the two-phase region is a failure possessed by any analytical equation of state.

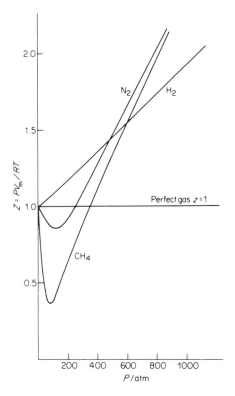

Figure 3.9. The compression factor as a function of pressure at $T = 273$ K. (Data from E. P. Bartlett *et al.*, *J. Amer. Chem. Soc.*, **52**, 1363 (1930) and H. M. Kvalnes and V. L. Gaddy, *J. Amer. Chem. Soc.*, **53**, 394 (1931))

$$\frac{\varepsilon_A}{kT_A} = \frac{\varepsilon_B}{kT_B} \quad \text{or} \quad \frac{T_A}{T_B} = \frac{\varepsilon_A}{\varepsilon_B}. \tag{3.56}$$

Likewise as σ is a scaling parameter for the distance between molecules and volume is of dimensions distance cubed, then gas A in volume V_A should have similar pair interactions to gas B in volume V_B if

$$\frac{V_{A,m}}{V_{B,m}} = \left(\frac{\sigma_A}{\sigma_B}\right)^3. \tag{3.57}$$

As pressure has the dimensions RT/V we then expect the ratio of the pressures to be

$$\frac{P_A}{P_B} = \left(\frac{\varepsilon_A}{\varepsilon_B}\right)\left(\frac{\sigma_B}{\sigma_A}\right)^3. \tag{3.58}$$

Let us put this another way; if temperature is measured in units of ε/k, volume in units of σ^3 and pressure in units of ε/σ^3 then the two fluids should obey the same equation of state

66

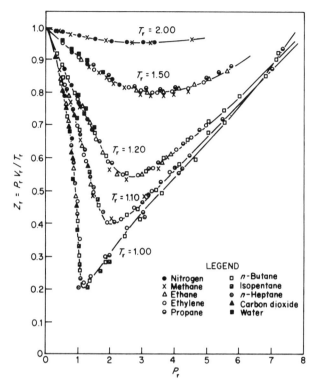

Figure 3.10. Reduced compressibility factor Z_r as a function of reduced temperature and pressure for a variety of gases. (From Gouq-Jen Su, *Ind. Eng. Chem.*, **38**, 803 (1946))

$$\phi(P^*, V^*, T^*) = 0, \qquad (3.59)$$

where

$$P^* = \sigma^3 P/\varepsilon, \quad V^* = V_m/\sigma^3, \quad T^* = kT/\varepsilon, \qquad (3.60)$$

are dimensionless variables. In particular, they should have the same critical constants (P_c^*, V_c^*, T_c^*). Because

$$P^*/P_c^* = P/P_c = P_r, \text{ etc.} \qquad (3.61)$$

the two fluids should obey the same reduced equation of state with variables (P_r, V_r, T_r).

The above argument can be generalized for any intermolecular potential in which there is a single energy-scaling factor and a single distance-scaling factor. We can also invert the argument: if an empirical equation of state in the reduced variables (P_r, V_r, T_r) can be found which holds for a number of different fluids then we deduce that the pair potentials of the molecules in all these fluids have the same functional form

$$V = \varepsilon f(R/\sigma), \qquad (3.62)$$

where ε and σ are energy and distance scale factors, respectively. Such fluids are said to belong to the same conformal family.

One consequence of the law of corresponding states is that the boiling point of a fluid at a fixed reduced pressure is a universal fraction of the critical temperature. Because the critical pressures of most fluids are similar (~ 50 atm) and boiling points on an absolute scale are relatively insensitive to pressure, we conclude that the normal boiling point ($P = 1$ atm) is a fixed fraction of the critical temperature, found experimentally to be approximately $2/3$.

A further generalization is that other thermodynamic properties of conformal families such as enthalpy, entropy and heat capacity can be given common functional forms when expressed in terms of the reduced variables. Figure 3.11 shows such a graph for heat capacity. This was prepared by Hougen and Watson by averaging the data for H_2, N_2, CO, NH_3, CH_4, C_3H_8 and C_5H_{12}.

The Virial Expansion

We showed in Section 3.3 that with some approximations one can derive a relationship (3.24) between the pressure of a gas and an integral over a function of the intermolecular potential. Knowing the pair potential U_{12} it is a straightforward matter to calculate the relevant integral (3.19) and hence

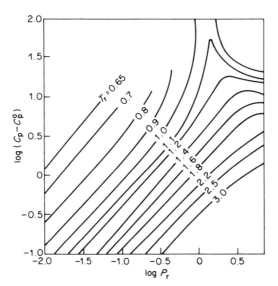

Figure 3.11. The difference between the heat capacity of a gas (C_p) and its perfect gas value (C_p^0) as a function of reduced pressure and reduced temperature (logarithmic scale). (O. A. Hougen and K. M. Watson, *Chemical Process Principles*, Part II, Wiley, London and New York (1947))

68

to deduce the pressure. It is, however, more common to make the connection in the reverse direction, that is, to take experimentally determined pressures and from them to deduce the pair potential. The analysis is not as simple as it might appear because the available data are relatively insensitive to certain features of the potential.

This is illustrated in Figure 3.12, which shows the pair potential for argon and the function f_{12} at 87 K. The b integral (3.19) is an integral over f_{12} weighted by R^2, and it is clear from the figure that the main contribution to the integral comes from the region of the potential around the minimum. The integral will be relatively insensitive to the short and very long-range parts of the potential. Indeed, the only feature of the repulsive branch of the potential that is likely to be deduced with reasonable accuracy from the experimental data is the point at which U_{12} crosses the R-axis (the parameter σ in Figure 2.8).

The most convenient way of making the connection between (P, V, T)-data and the intermolecular potential is through an equation of state for gases known as the virial equation (first used by Kammerlingh-Onnes and

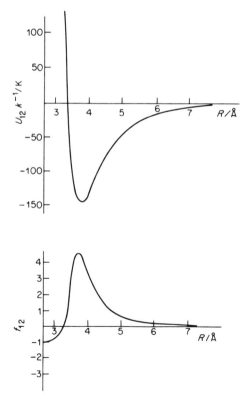

Figure 3.12. The pair potential for argon and the resulting function $f_{12} = \exp(-U_{12}/kT) - 1$, at 87 K

Keesom in 1912). It is in the form of an expansion of the pressure in powers of V^{-1}.

$$P = \frac{RT}{V_m}\left[1 + \frac{B(T)}{V_m} + \frac{C(T)}{V_m^2} = \ldots\right]. \tag{3.63}$$

$B(T)$ is known as the second virial coefficient, $C(T)$ is the third virial coefficient, etc; it is important to note that they are functions of temperature. The relationship between the b integral and $B(T)$ can be obtained by equating the terms of order $(V_m)^2$ in 3.24 and 3.63.

$$B(T) = 2\pi \mathcal{N}b = 2\pi \mathcal{N}\int (1 - e^{-U_{12}/kT})R^2\, dR \tag{3.64}$$

From the cluster expansion of the configurational integral which was described in Section 3.3, it can be shown that the third virial coefficient depends on both the pair potential and the three-body potential.

Figure 3.13 shows the second virial coefficient of argon as a function of temperature. $B(T)$ is negative until rather high temperatures are reached. The change in sign occurs at the Boyle temperature (412 K for argon) and at this temperature the gas is, to a good approximation, behaving as a perfect gas. At the Boyle temperature the effects of the attractive and the repulsive parts of the potential just cancel in the b integral.

Table 3.3 lists some Ar_2 pair potentials that have been optimized by a least-squares procedure to $B(T)$. The L-J, square-well and (exp − 6) potentials were described in Chapter 2 (see Figure 2.8). The Kihara potential is a modified L-J potential with an extra parameter to improve the repulsive branch. The L-J potential is the only two-parameter potential in

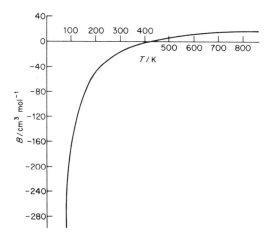

Figure 3.13. Second virial coefficient of argon as a function of temperature. A smooth curve has been drawn through data from the following references: R. D. Weir et al., *Trans. Faraday Soc.*, **63**, 1320 (1967); E. Whalley et al., *Can. J. Chem.*, **31**, 722 (1953). The first of these papers has a very good description of the experimental technique

Table 3.3. Parameters of argon pair potentials which have been optimized to second virial coefficients

Potential	$\sigma/\text{Å}$	$(\varepsilon/k)/\text{K}$	r.m.s./$\text{cm}^3\,\text{mol}^{-1}$
L-J	3.504	118	2.02
Kihara	3.314	147	0.51
Square-well	3.067	93	0.49
(exp − 6)	3.302	152	0.54
Parson *et al.*	3.354	141	−

A. E. Sherwood and J. M. Prausnitz, *J. Chem. Phys.*, **41**, 429 (1964); J. M. Parson, P. E. Siska and Y. T. Lee, *J. Chem. Phys.*, **56**, 1511 (1972). r.m.s. is the root mean square error.

the list and it is significantly poorer than the others in fitting $B(T)$. The square-well potential gives a surprisingly good fit to the data although in both the depth of the well and the value of σ it is very poor when compared with the very accurate potential of Parson and co-workers; this is a complex potential deduced from atomic beam scattering that also fits $B(T)$ with high accuracy.

The above example shows that one must be cautious in accepting a potential which is based solely on second virial coefficient data. Different types of experiment probe different regions of the potential, and hence a combination of data from different experiments is a much sterner test of a potential than are extensive data from a single type of experiment. Some of the most useful data for providing information about pair potentials comes from gas-transport properties, and this is our next topic for examination.

3.7 TRANSPORT PROPERTIES

Up to this point we have been concerned mainly with the properties of fluids in equilibrium, a state of equilibrium being characterized by a uniformity of properties throughout the volume of the system and a constancy of these properties with time. We shall now discuss briefly some important properties of non-equilibrium fluids, properties which are technologically important and have a bearing on the molecular interpretation of fluid behaviour.

Although the fundamentals of non-equilibrium statistical mechanisms and thermodynamics are not so well established as their equilibrium counterparts that is not through lack of attention by scientists. The kinetic theory of gases was developed in the second half of the nineteenth century and culminated in the Boltzmann transport equation, which describes the time evolution of the phase-space probability density $f^{(N)}(\boldsymbol{p}, \boldsymbol{q})$ (3.25) for an N particle system in which collisions are occurring. Boltzmann's great achievement was to show that this kinetic equation incorporates the fundamental irreversibility of many-particle systems which is contained in the principle that spontaneous change occurs in the direction of increasing entropy.

The Boltzmann equation can be solved for gases by assuming that only binary collisions occur between molecules. Transport properties such as viscosity can then be related to parameters of these collisions (e.g. angles of deflection) which can be calculated from the intermolecular potential. As for the treatment of virial coefficients, we can proceed in the opposite direction and deduce potentials from transport properties.[†] For dense fluids, however, there are still substantial problems in obtaining analytical solutions of the transport equations and their mathematical complexity makes them inappropriate for detailed discussion in this book.

An alternative approach to non-equilibrium processes which is based upon an extension of thermodynamic principles is known as the phenomenological approach. This term implies that the equations which are used are based upon experimental observation rather than on any underlying assumptions about the molecular interactions in the system. The approach was formalized by Onsager in 1931 although many of the equations and general principles were known much earlier.

The essential pattern of the phenomenological equations is based on the observation that the rate of flow of some property, which we shall represent by the vector flux J, is proportional to a generalized force, F, which is in turn derived from the gradient of some property, ϕ. Thus we can write

$$J \propto F, \tag{3.65}$$

or

$$J \propto \text{grad } \phi. \tag{3.66}$$

The proportionality factors are called the phenomenological coefficients. These equations can be extended to situations in which there is more than one flux (e.g. matter and heat) and it was this generalization that was the key idea in Onsager's theory.

Let us take an example: the flux of thermal energy is proportional to the temperature gradient in the system. Thus the z-component of the flux can be written

$$J_z(\text{thermal energy}) = -\kappa(\mathrm{d}T/\mathrm{d}z), \tag{3.67}$$

which is known as Fourier's law. The negative sign is inserted because a flux towards increasing z (J_z positive) will occur when there is a negative temperature gradient. The constant κ is called the coefficient of thermal conductivity.

A second example is the flow of matter down a concentration gradient. The relevant equation is

$$J_z(\text{matter}) = -D(\mathrm{d}C/\mathrm{d}z), \tag{3.68}$$

[†]S. Chapman and T. A. Cowling, *The Mathematical Theory of Non-uniform Gases*, C.U.P. (1939).

where C is the concentration and D is called the diffusion coefficient. Equation 3.68 is known as Fick's first law of diffusion (his second law relates to a differential equation for C but it has no relevance to our present discussion).

Viscosity

Finally we turn to the transport property which has probably most importance for pure liquids and that is the viscosity. Viscous forces appear only when adjacent parts of a fluid are moving with different velocities. Thus in the flow of liquid down a pipe the liquid layer which is actually in contact with the pipe is stationary and the axial region is moving with maximum velocity. Common experience is that a high-viscosity liquid like treacle will flow more slowly than a low viscosity liquid like water for a given applied pressure difference.

The above experimental arrangement provides a simple method for measuring liquid viscosities. Poiseuille's formula for the flow of liquid in a capillary of radius R and length l is

$$\frac{dV}{dt} = (P_1^2 - P_2^2)\pi R^4 / 16 l \eta P. \tag{3.69}$$

V is the volume flowing, P_1 and P_2 are the pressures at each end of the tube, P is the pressure at which the volume is measured and η is the coefficient of viscosity. The rate of flow of liquid is measured as a function of the applied pressure and interpreted by the above formula.

For the basic transport equation for viscosity we require the flux of some quantity which is maintained by a velocity gradient. To see what this quantity is we take a theoretically simpler arrangement than the flow down a capillary. Figure 3.14 shows two parallel plates with liquid between, the bottom plate being stationary and the top plate moving with constant velocity v_x^0. For gases and most liquids the force per unit area of the top

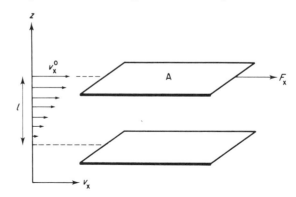

Figure 3.14. The laminar flow of a fluid between a moving upper plate and a fixed lower plate

plate required to maintain this constant velocity is found to be proportional to v_x^0 and such fluids are called Newtonian fluids. Non-Newtonian behaviour is often exhibited by very viscous fluids, typically polymer solutions, and we will leave a discussion of these to Chapter 11.

As the bottom layer of liquid is stationary and the top layer is moving at v_x^0 we must ask how the intermediate layers are moving. The simplest assumption, and one that can be confirmed by experiment, is that the velocity is changing linearly with distance between the plates as indicated in Figure 3.14. The fluid is then said to be in a state of laminar flow characterized by a velocity gradient

$$\frac{dv_x}{dz} = \frac{v_x^0}{l}. \tag{3.70}$$

The experimental situation can then be summarized in the equation

$$\frac{F_x}{A} = \frac{\eta v_x^0}{l} = \eta \left(\frac{dv_x}{dz} \right), \tag{3.71}$$

where η is the coefficient of shear viscosity; its SI unit is $kg\,m^{-1}\,s^{-1}$.

Let us now imagine the fluid to be divided into layers of molecular thickness. There will be a constant exchange of molecules between layers and these molecules will be carrying momentum in the x direction which will tend to bring the two layers to the same speed; the molecules from the faster layer will speed up the slower and these from the slower layer will slow down the faster. It is the applied force F_x which conteracts this drag. We therefore see that the flux which is associated with the velocity gradient is a flux of momentum in the x direction. We can therefore write an expression analogous to 3.67 and 3.68 as follows

$$J_z(x \text{ momentum}) \propto \left(\frac{dv_x}{dz} \right), \tag{3.72}$$

where the proportionality constant will be the coefficient of viscosity with appropriate units.

As we have already indicated, the transport coefficients can be calculated on a molecular basis from the kinetic theory of gases, the key parameter being the average distance travelled by molecules between collisions (the mean free path). For liquids such a theory is inapplicable as the concept of distinct binary collisions is inappropriate to a liquid.

From gas kinetic theory it is predicted that viscosity is independent of pressure and proportional to the square root of the absolute temperature. These predictions are confirmed by experiment. For liquids, however, the experimental behaviour is quite different. Liquid viscosities have an exponential dependence on temperature as described by the formula

$$\eta = A \exp (\Delta E_{vis}/RT). \tag{3.73}$$

Figure 3.15 shows a graph of $\log \eta$ against $10^3/T$ for CCl_4.

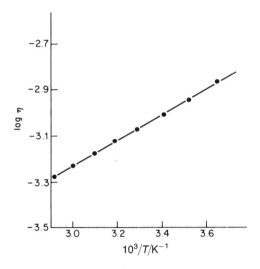

Figure 3.15. A graph of $\log(\eta/\text{kg m}^{-1}\text{s}^{-1})$ against T^{-1} for CCl_4

It has been customary to interpret ΔE_{vis} as an activation barrier that the molecules must overcome before the liquid can flow. Hildebrand has argued against this interpretation and has pointed out that η is an extremely sensitive function of molar volume, liquids flowing very freely when expanded only a few per cent above their intrinsic volume. He suggested that fluidity (measured by η^{-1}) should be proportional to the ratio of free volume to occupied volume (the volume of the molecules) in the liquid. If

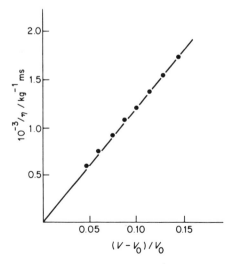

Figure 3.16. The relationship between fluidity $(1/\eta)$ and the ratio of free to filled volume for CCl_4. (Data from J. Hildebrand, *J.C.S. Faraday Disc.*, **66**, 151 (1978))

V_0 is the occupied volume and V the actual volume, then Hildebrand's relationship is

$$\frac{1}{\eta} = B\left(\frac{V - V_0}{V_0}\right), \tag{3.74}$$

where B and V_0 are parameters deduced from the straight line graph $1/\eta$ against V. Figure 3.16 shows for CCl_4 that this interpretation is supported by experimental data. Hildebrand attributes the empirical relationship 3.73 to the temperature dependence of the ratio $(V - V_0)/V_0$.

APPENDIX 3.1

The integral in 3.17 can be reduced further by choosing as independent coordinates, (X, Y, Z), the position of the centre of mass of the two particles, and (R, θ, φ) the polar coordinates of particle 2 relative to particle 1. The volume element in these coordinates is

$$dX \, dY \, dZ R^2 \, dR \sin \theta \, d\theta \, d\varphi. \tag{A3.1}$$

Integration over X, Y and Z gives a factor V and integration over θ and φ gives a factor 4π. We therefore have

$$\iint f_{12} \, dq_1 \, dq_2 = 4\pi V \int f_{12}(R) R^2 \, dR. \tag{A3.2}$$

For molecules the situation is more complicated and a detailed treatment can be found elsewhere.†

APPENDIX 3.2

The Metropolis method for applying the Monte Carlo procedure to the computer simulation of a liquid is as follows:

(1) Suppose a random configuration, i, has been constructed with energy U_i.

(2) One particle in this configuration is selected either serially (1 to N) or randomly, and by choosing three random numbers $(\zeta_1, \zeta_2, \zeta_3)$ in the interval 0 to 1 this particle is displaced along the three cartesian axes by amounts

$$x = (2\zeta_1 - 1)k, \quad y = (2\zeta_2 - 1)k, \quad z = (2\zeta_3 - 1)k. \tag{A3.3}$$

A new energy U_{i+1} is calculated.

(3) If U_{i+1} is less than U_i the new configuration is added to the set and this configuration becomes the starting point for a further repetition of step 2.

(4) If U_{i+1} is greater than U_i the new configuration is not immediately selected but is judged by choosing a further random number η in the range 0 to 1. If

$$\eta \leqslant \exp\left(-(U_{i+1} - U_i)/kT\right), \tag{A3.4}$$

†J. O. Hirschfelder, C. F. Curtis and R. B. Bird, *Molecular Theory of Gases and Liquids*, Wiley, London and New York (1964).

then configuration i + 1 is added to the set and it becomes the starting point for step 2. If, however,

$$\eta > \exp\left(-(U_{i+1} - U_i)/kT\right), \tag{A3.5}$$

the configuration is rejected and the previous configuration i is counted again.

It can be seen that this procedure generates a set of configurations which are generally low in energy, and it can be shown that they occur with a Boltzmann probability. The parameter k, in A3.3 is quite important. If it is too large then many of the new configurations which are generated will be rejected. If it is too small then it will take a long time to generate a representative set. At typical liquid densities a value of k which is approximately one-fifth of the distance between nearest neighbours has been found to be a reasonable compromise, and typically about half of the generated configurations are rejected by the criterion A3.5. Averages are normally evaluated from a total of 10^4–10^5 configurations.

APPENDIX 3.3

To prove Maxwell's equal area rule (page 64), we start with the formula for integration by parts

$$\int_{V_1}^{V_2} P \, dV = |PV|_{V_1}^{V_2} - \int_{V_1}^{V_2} V \, dP. \tag{A3.6}$$

Let V_1 and V_2 be the molar volumes of the liquid and gas respectively, when they are in equilibrium at the same temperature and pressure. For an isothermal change $V \, dP$ is equal to dG the change in Gibbs free energy. As liquid and gas in equilibrium have the same molar free energy, dG must be zero, hence A3.6 becomes

$$\int_{V_1}^{V_2} P \, dV = P(V_2 - V_1). \tag{A3.7}$$

The left-hand side of this equation is equal to the area under the van der Waals isotherm and the right-hand side is the area under the straight line joining the points (P, V_1) and (P, V_2). These two areas will be equal only if the areas in Figure 3.8 are equal.

Chapter 4

Thermodynamic Properties of Pure Liquids

4.1 THERMODYNAMICS OF PHASE EQUILIBRIA

In this chapter we examine the thermodynamics of pure liquids; i.e. systems with a single chemical component, placing particular emphasis on the solid–liquid and liquid–gas phase equilibria. Other aspects of pure liquids such as liquid surfaces will be dealt with in Chapters 12 and 13.

We write this chapter on the assumption that the reader is familiar with the definitions of the main thermodynamic variables, with the equations linking these variables, and with the criteria for equilibrium. The following introductory remarks are meant to refresh the memory but not to be comprehensive. The necessary criterion for spontaneous change in a closed system maintained at constant temperature and pressure is that there shall be a decrease in the Gibbs free energy of the system, which is related to the enthalpy and entropy of the system by

$$G = H - TS. \tag{4.1}$$

The rate at which spontaneous change occurs depends on mechanistic (kinetic) factors which we do not cover in this book. For example, diamond is thermodynamically unstable relative to graphite at room temperatures, but is kinetically completely stable. Likewise, if the temperature of a liquid is taken rapidly below its freezing point, the liquid may form a glass rather than the thermodynamically more stable crystalline solid.

The Chemical Potential

G, H and S are extensive quantities, that is, their values are proportional to the amount of material in the system. For single-component systems their

78

values per mole are given the symbols G_m, H_m and S_m. The molar Gibbs free energy is an extremely important quantity and it is given another symbol, μ, and name, the chemical potential. For multicomponent systems the concept of molar quantities must be extended, and this topic is dealt with in Chapter 7. We only emphasize at this point that μ has greater generality than G_m, but that it is equal to G_m for a pure substance.

For thermal equilibrium between phases, each phase must be at the same temperature. For mechanical equilibrium between phases (in the absence of external fields), each phase must be at the same pressure. If these two conditions are met, then the criterion for total equilibrium between phases is that their chemical potentials shall be equal, i.e., there is also physiochemical equilibrium. We note that μ is like T and P in being an intensive property of the system, and is also like T and P in providing a criterion of equilibrium.

In mechanical systems the direction of spontaneous change is that which leads to a decrease in the mechanical potential. Thus a ball rolls downhill because its potential (gh, where g is the acceleration due to gravity and h is its height) decreases. In chemical systems the direction of spontaneous change is that which leads to a decrease in μ, the chemical potential.

The most stable phase of a pure substance at a particular temperature and pressure is the one with the lowest chemical potential. At low temperatures μ will be dominated by H_m and ordered structures with strong interactions between the molecules (i.e. a solid) will be favoured. At high temperatures μ will be dominated by the factor TS_m and a system with the greatest entropy or disorder will be favoured, which is the gas. The liquid state is therefore an intermediate situation in which neither H_m nor TS_m predominates.

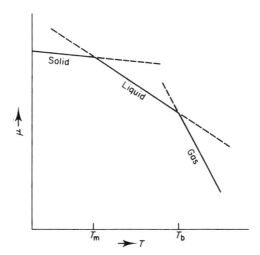

Figure 4.1. Schematic representation of the chemical potential for solid, liquid and gas as a function of temperature (fixed pressure)

As we shall demonstrate, μ decreases with increasing temperature, and Figure 4.1 shows schematically the variation of μ with temperature for typical solid, liquid and gas phases. The dashed lines show the continuation of the functions into temperature ranges where the phase is not the most stable. The melting point is the temperature at which the intersection of $\mu(s)$ and $\mu(1)$ occurs; the boiling point is at the intersection of $\mu(1)$ and $\mu(g)$.

The simplicity of Figure 4.1 disguises its importance. Let us take one example. Suppose we make some change to the system which lowers the chemical potential of the liquid state but leaves $\mu(s)$ and $\mu(g)$ unchanged. Such a change will have the effect of *lowering* the melting point and *raising* the boiling point of the system. We will see in a later chapter that this can be achieved by dissolving an involatile solute in the liquid.

As the stability of a phase depends on temperature and pressure we are interested in knowing how the Gibbs free energy varies with temperature and pressure. The key to this information is the equation

$$dG = V \, dP - S \, dT, \tag{4.2}$$

which tells us how small changes in pressure and temperature lead to a small change in G. dG is called an exact differential because a finite change in G depends only on the difference between its values in the initial and final states. dG does not depend on the path (in P, T space) by which this change is carried out and, in particular, on whether the change is carried out reversibly or not. From 4.2 the partial derivatives of G are given by

$$\left(\frac{\partial G}{\partial P} \right)_T = V, \tag{4.3}$$

and

$$\left(\frac{\partial G}{\partial T} \right)_P = -S. \tag{4.4}$$

Similar expressions hold also for the corresponding molar quantities, hence we can write

$$\left(\frac{\partial \mu}{\partial P} \right)_T = V_m \quad \text{and} \quad \left(\frac{\partial \mu}{\partial T} \right)_P = -S_m. \tag{4.5}$$

The function S has a special place amongst the extensive quantities of thermodynamics in that its absolute value can be determined and not, as for H and G, just relative values. The reason for this is contained in the third law of thermodynamics which can be stated as, 'all perfect crystals at absolute zero have zero entropy'. From this reference point absolute entropies can be deduced at other temperatures, and these entropies are always positive. We, therefore, make the important deduction from 4.5 that at constant pressure the chemical potential of a system always decreases with increasing temperature, and that the rate of decrease is greatest for those states with the greatest entropy. Figure 4.1 is consistent with these conclusions.

The difference in the chemical potentials of two states of a single component at the same temperature and pressure is, from 4.1,

$$\Delta\mu = \Delta H_m - T\Delta S_m = \Delta U_m + P\Delta V_m - T\Delta S_m. \tag{4.6}$$

For two phases in equilibrium $\Delta\mu$ is zero so that

$$\Delta H_m = T\Delta S_m. \tag{4.7}$$

Both melting and vaporization are endothermic processes, hence they are accompanied by an increase in both enthalpy and entropy. The molar volume will also generally change and so will ΔU_m.

If V_m and S_m are discontinuous at a phase change, it follows from 4.5 that $(\partial\mu/\partial P)_T$ and $(\partial\mu/\partial T)_P$ are also discontinuous. This is illustrated in Figure 4.1 when the slope of μ is discontinuous at the melting and boiling points. A phase change having these thermodynamic characteristics is called a first-order phase transition.

There are other phase changes in which there is no change in volume or entropy although there will be a discontinuity in the derivatives of these quantities with temperature or pressure at the phase change. In these cases $(\partial\mu/\partial P)_T$ and $(\partial\mu/\partial T)_P$ are continuous at the phase transition but the second derivatives of μ are discontinuous. Such transitions are called second-order phase changes; paramagnetic–ferromagnetic changes in a solid are of this type. Figure 4.2 summarizes the temperature behaviour of the thermodynamic quantities we have referred to as first-order and second-order phase transitions.

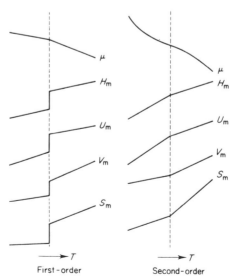

Figure 4.2. Changes in thermodynamic quantities at first- and second-order phase transitions

Effect of Pressure and Temperature

From expression 4.3 we can deduce the effect of pressure on the melting and boiling points. On passing from liquid to gas there is always a large increase in volume so that an increase in pressure will raise $\mu(g)$ much more than $\mu(1)$. From Figure 4.3 it is clear that this will increase the temperature at which the liquid boils. The change is consistent with le Chatelier's principle that a system in equilibrium tends to move in a direction which minimizes any outside perturbation. That is, an increase in pressure will move the system towards the phase with the smaller volume.

In general, the solid to liquid phase change is also accompanied by an increase in volume so that an increase in pressure raises the melting point as in Figure 4.3. However, in a few cases, the most important being water, there is a decrease in volume on melting so that for these systems an increase in pressure will lower the melting point. It is common knowledge that applying sufficient pressure to ice will cause it to melt.

We can derive from 4.2 an expression which determines the effect of pressure on the temperature at which a phase change occurs. Suppose we make changes in P and T in such a way that two phases α and β remain in equilibrium. We then have for one mole of each phase

$$d\mu(\alpha) = V_m(\alpha)\,dP - S_m(\alpha)\,dT, \tag{4.8}$$

and

$$d\mu(\beta) = V_m(\beta)\,dP - S_m(\beta)\,dT, \tag{4.9}$$

and, if equilibrium is maintained,

$$\mu(\alpha) = \mu(\beta) \quad \text{and} \quad d\mu(\alpha) = d\mu(\beta). \tag{4.10}$$

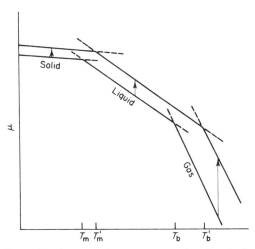

Figure 4.3. The effect of an increase in pressure on the chemical potentials of solid, liquid and gas, and the resulting increase in the melting and boiling temperatures

We can therefore write

$$V_m(\alpha)\,dP - S_m(\alpha)\,dT = V_m(\beta)\,dP - S_m(\beta)\,dT, \tag{4.11}$$

or

$$\Delta V_m\,dP = \Delta S_m\,dT, \tag{4.12}$$

where Δ symbolises the difference between the molar quantities for the two phases. Equation 4.12 can be rearranged to give

$$\frac{dP}{dT} = \frac{\Delta S_m}{\Delta V_m} = \frac{\Delta H_m}{T\Delta V_m}, \tag{4.13}$$

which is called the Clapeyron equation. This equation is exact and applies to any phase change.

Both ΔH_m and ΔV_m are always positive for the solid $\rightarrow$ gas and liquid $\rightarrow$ gas phase changes, hence we conclude that the corresponding boundary lines in the P, T phase diagram (which are the lines along which equilibrium of the two phases is maintained) always have positive slope. This is also true in most cases for the solid $\rightarrow$ liquid line with water again being an exception (see Figure 1.2).

For the solid $\rightarrow$ liquid phase change we can make use of the fact that ΔH_m and ΔV_m do not vary appreciably with pressure or temperature, and integrate 4.13 to give

$$P = P_0 + \left(\frac{\Delta H_{melt}}{\Delta V_{melt}}\right)\ln(T/T_0), \tag{4.14}$$

where (P_0, T_0) is some known point on the phase boundary line. This expression will, however, be accurate only over a short temperature range.

For the solid $\rightarrow$ gas and liquid $\rightarrow$ gas transitions we cannot make this approximation, because the volume of a gas depends on pressure and temperature. An alternative aproach is to use the fact that the gas has a much larger molar volume than the liquid or solid hence, assuming that the vapour behaves as a perfect gas

$$\Delta V_m \approx V_m(g) = RT/P. \tag{4.15}$$

On substitution in 4.13 this gives us an equation known as the Clausius–Clapeyron equation

$$\frac{dP}{dT} = \frac{P\Delta H_m}{RT^2}, \tag{4.16}$$

which can be written in the alternative form

$$\frac{d(\ln P)}{d(1/T)} = -\frac{\Delta H_m}{R} \tag{4.17}$$

where ΔH_m is the enthalpy change for the solid $\rightarrow$ gas or liquid $\rightarrow$ gas transition.

If one measures the vapour pressure of a liquid as a function of temperature (the vapour pressure is the pressure at which the liquid and vapour are in equilibrium at a specified temperature) then it is generally found that a graph of ln P against $1/T$ is nearly linear over a wide range of temperature. This is illustrated for water in Figure 4.4 which shows the temperature range from the ice point to the critical point.

Under these circumstances one can deduce an empirical expression for the vapour pressure

$$\ln (P/P_0) = -A\left(\frac{1}{T} - \frac{1}{T_0}\right), \tag{4.18}$$

which is clearly valuable for interpolation purposes. However, the temptation to compare 4.18 with an integrated form of 4.17 which assumes that ΔH_m is independent of temperature

$$\ln (P/P_0) = -\frac{\Delta H_m}{R}\left(\frac{1}{T} - \frac{1}{T_0}\right), \tag{4.19}$$

must be resisted because ΔH_m is far from being constant over such a wide temperature range. In fact ΔH_m must be zero at the critical point for there the properties of liquid and vapour coincide.† We know also that the assumptions made in deriving 4.16 are invalid close to the critical point

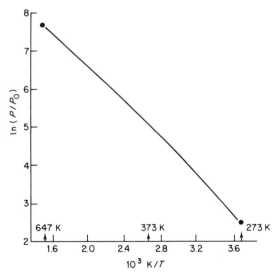

Figure 4.4. A graph of ln P against T^{-1} for water from the critical point to the ice point

---

†Note that the ratio $\Delta S_m/\Delta V_m$, which appears in 4.13, is indeterminate at the critical point, although (dP/dT) is certainly finite. Application of l'Hôspital's rule does not resolve the matter as differentiating still leaves an indeterminate ratio. We have commented earlier (page 60) on the strange behaviour of some thermodynamic parameters at the critical point.

84

because the gas does not then have a much larger molar volume than the liquid; they become equal at this point. In short, expression 4.18 can be used to deduce the enthalpy of vaporization only in the low pressure regions where the Clausius–Clapeyron assumptions are valid. The enthalpy of vaporization of water as a function of temperature is shown in Figure 4.5.

Entropy Changes on Melting and Vaporization

The entropy change on melting a solid is generally rather small but varied in its magnitude. Values usually occur in the range 10–$40\,\mathrm{J\,K^{-1}\,mol^{-1}}$. There are several contributions to this. In the first place there is the contribution which is associated with the fact that exchange of molecules between lattice sites in a solid does not occur whereas exchange between positions in a liquid does. We mentioned this point in section 3.3 when discussing the factor $N!$ that occurs in expression 3.5. The contribution from this is about $8\,\mathrm{J\,K^{-1}\,mol^{-1}}$ and the inert gases have entropies of melting which are close to this value. For polyatomic molecules the main contribution arises from the change of lattice vibrations of the solid to restricted translations and rotations in the liquid.

For the liquid to gas transition the entropy change is larger and for most molecules it is approximately $85\,\mathrm{J\,K^{-1}\,mol^{-1}}$. This constancy is known as Trouton's rule and it can be understood from the fact that gases have a much greater degree of disorder than liquids, hence differences in the amount of order possessed by one liquid and another are small compared with the change on going to the gas. However, as Hildebrand has emphasized, the gases produced at the normal boiling points of liquids do not all have the same molar volumes because the boiling points differ. As the entropy of a gas depends on volume and temperature we should expect to find greater constancy for the entropies of vaporization at constant molar volume and this is generally true.

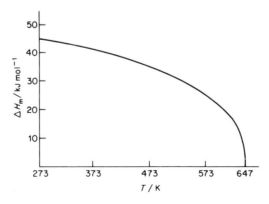

Figure 4.5. The enthalpy of vaporization of water as a function of temperature

From the Maxwell relation

$$\left(\frac{\partial S}{\partial V}\right)_T = \left(\frac{\partial P}{\partial T}\right)_V \tag{4.20}$$

we can see that for a perfect gas

$$\left(\frac{\partial S}{\partial V}\right)_T = \frac{R}{V} \tag{4.21}$$

hence the dependence of entropy on volume for a perfect gas at constant temperature is

$$S_m(V_2) - S_m(V_1) = R \ln (V_2/V_1) \tag{4.22}$$

Thus if we have two liquids whose boiling points differ by a factor of two (in degrees Kelvin) then their molar volumes of gas at the boiling points will differ by a factor of two and the contribution to the entropy difference is $R \ln 2 = 5.8 \ \mathrm{J K^{-1} mol^{-1}}$.

The fact that such a difference is small compared with the total entropy change explains why Trouton's rule is quite good even without Hildebrand's improvement.

If the entropy of vaporization differs appreciably from the Trouton constant then that indicates that there is some unusual behaviour in the liquid or gaseous states. For example, water and ethanol can be seen from Table 4.1 to have larger than normal entropies of vaporization and that must indicate that their liquids have a greater than normal degree of order, hence a lower than normal entropy. This can be attributed to a hydrogen-bonded structure within the liquid. In contrast ethanoic acid has a lower than normal entropy of vaporization, as do other aliphatic acids, and this indicates that some order is retained in the gas phase by virtue of dimerization.

Table 4.1. The boiling point T_b and entropy of vaporization (column (1)) for some liquids. Column (2) shows the entropy of vaporization at a constant molar volume of 49.56 dm³

		$\Delta S_m/\mathrm{J \ K^{-1} mol^{-1}}$	
	T_b/K	(1)	(2)
Hg	630	94.1	84.1
CH_4	112	82.8	94.1
nC_4H_{10}	272	82.0	93.3
cyclo-C_6H_{12}	354	84.9	92.9
C_6H_6	353	87.1	95.0
H_2O	373	109.0	145.6
C_2H_5OH	452	139.7	149.8
CH_3COOH	392	62.3	—

See J. H. Hildebrand, *J. C. S. Faraday Disc.*, **66** (1978); R. W. Hermsen and J. M. Prausnitz, *J. Chem. Phys.*, **34**, 1081 (1961).

4.2 THE CHEMICAL POTENTIAL OF A GAS

Although this book is mainly concerned with the liquid state we cannot omit a discussion of the chemical potential of the perfect and imperfect gas for two reasons. First, above the critical point gas and liquid become a single state of matter. Second, and more important for our purposes, the simplest way of determining the chemical potential of a liquid is to determine the chemical potential of the gas with which it is in equilibrium. An additional point is that our approach to non-ideality in a gas will be closely paralleled later in the book when we discuss non-ideal solutions.

If we integrate expressions 4.5 we can calculate the chemical potential at one value of temperature and pressure knowing its value at another. For example, we can write for constant T

$$\mu(P_2) - \mu(P_1) = \int_{P_1}^{P_2} V_m(P) \, dP. \tag{4.23}$$

For an ideal gas we know that the volume is inversely dependent on the pressure, and hence can arrive at the result

$$\mu(P_2) - \mu(P_1) = RT \int_{P_1}^{P_2} P^{-1} \, dP = RT \ln\left(\frac{P_2}{P_1}\right). \tag{4.24}$$

In establishing tables of reference for thermodynamic quantities it is convenient to adopt standard conditions of temperature and pressure and any other variables that might influence the system. By using expressions such as 4.24 one can then calculate the changes in the relevant quantity from this so-called standard state. For an ideal gas the standard state is defined to be at a pressure of 1 atm so that if we denote the standard chemical potential by $\mu^{\ominus}$, we have from 4.24

$$\mu(P) = \mu^{\ominus} + RT \ln(P/\text{atm}). \tag{4.25}$$

A standard temperature of 298 K is also normally adopted. The value of μ at other temperatures can be deduced from expression 4.5 if one knows the dependence of entropy on the temperature.

There is one mathematical feature of 4.25 that is worth noting. As $\ln(0) = -\infty$ it follows that the chemical potential of an ideal gas approaches $-\infty$ as the pressure tends to zero. As real gases all become ideal gases at low pressures it follows that the chemical potentials of real gases also approach $-\infty$ as the pressure approaches zero.

Fugacity

For real gases the integral in 4.23 must be evaluated from careful measurement of the pressure-dependence of the molar volume. However, as gas imperfections are not usually very large it is convenient to use an expression for the chemical potential of a real gas which shows its relationship to the ideal gas. To do this we define a function f called the fugacity, which has the dimensions of pressure, by

$$\mu(P) = \mu^{\ominus} + RT \ln(f/\text{atm}),\tag{4.26}$$

subject to the condition

$$\underset{P \to 0}{\text{Limit}}\left(\frac{f}{P}\right) = 1.\tag{4.27}$$

It must be emphasized that the fugacity is defined by both of these expressions; 4.26 alone is not enough as that just amounts to an arbitrary division of $\mu(P)$ into two terms.

The relationship between f and P can be expressed more clearly by defining a dimensionless fugacity coefficient γ by

$$f = \gamma P,\tag{4.28}$$

so that

$$\mu(P) = \mu^{\ominus} + RT \ln(P/\text{atm}) + RT \ln \gamma,\tag{4.29}$$

where

$$\underset{P \to 0}{\text{Limit}}\ \gamma = 1.\tag{4.30}$$

Although $\mu^{\ominus}$ is, from 4.26, the value of the chemical potential when $f = 1$ atm, this is of no help for its determination because we do not know the pressure at this fugacity. As gases differ in their deviation from ideality $\mu^{\ominus}$ would have to be determined at different pressures for different gases. However, $\mu^{\ominus}$ can be obtained from measurements of μ at low pressures where the gas can be taken as ideal.

$$\mu^{\ominus} = \mu(P_{\text{low}}) - RT \ln f_{\text{low}} = \mu(P_{\text{low}}) - RT \ln P_{\text{low}}\tag{4.31}$$

and values of f at high pressures are obtained from 4.23 by integrating from a low pressure ($P_1 = P_{\text{low}}$) to the pressure of interest ($P_2 = P$). From 4.23 and 4.26, we have

$$\mu(P) - \mu(P_{\text{low}}) = \int_{P_{\text{low}}}^{P} V_m(P)\,dP$$

$$\ln f = \ln P_{\text{low}} + \frac{1}{RT}\int_{P_{\text{low}}}^{P} V_m(P)\,dP.\tag{4.32}$$

Figure 4.6 shows the dependence of γ on pressure for N_2 at 273 K. At pressures below about 300 atm γ is less than one but the gas is almost ideal. Above this pressure γ, and hence the non-ideality, increases rapidly. These two regions parallel the behaviour of the compression factor Z (see Figure 3.9) and reflect the importance of the a and b parameters of the van der Waals equation (3.44). Different gases have almost the same fugacity coefficient when they are at the same reduced temperature and pressure.

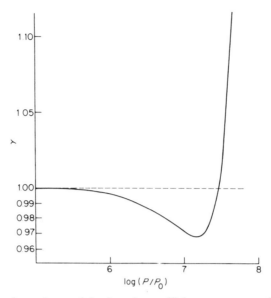

Figure 4.6. The dependence of the fugacity coefficient on pressure for N_2 at 273 K

Let us now ask whether there is a physical definition of the standard state of a real gas. If we take 4.29 as our starting point then we note that if γ measures the whole deviation from ideality $\mu^{\ominus}$ must be taken as the chemical potential of an ideal gas. In other words the standard state of a real gas is a hypothetical state in which the gas is behaving ideally and in which the fugacity is one atmosphere—that is, the pressure is one atmosphere.

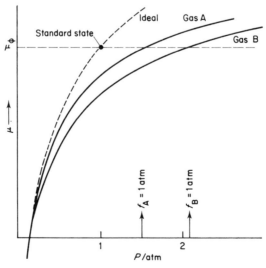

Figure 4.7. The chemical potentials of two real gases and their common standard state

Thus all real gases have the same standard state which is a hypothetical ideal gas at $P = 1$ atm. Figure 4.7 illustrates this definition. It will be seen later in the book that a similar approach is made to the standard states of solutions.

4.3 SINGLE-COMPONENT PHASE DIAGRAMS

Our first look at phase diagrams in Section 1.2 had the objective of introducing terms like the triple point and critical point and emphasizing the continuity of the fluid phase above the critical point. As we have now established the concept of the equation of state (Section 3.6) and discussed the thermodynamics of phase transitions (Section 4.1) we can return to the single-component phase diagram and examined it in more detail.

The general equation of state for a single component (3.41) can be represented by a surface in three-dimensional space with pressure, temperature and molar volume as cartesian axes. A typical surface is shown in Figure 4.8. The single-phase regions of the surface are curved in the three dimensions, the two-phase regions, e.g. liquid + gas, have zero slope along the volume axis because for a specified temperature the pressure is constant whatever the proportions of the two phases. The triple-point line shows the conditions of fixed pressure and fixed temperature if all three phases are in equilibrium: the volume depends on what proportions of gas, liquid and solid arc in the equilibrium mixture.

The (P, T)-phase diagram shown in Figure 1.1 is a projection of the

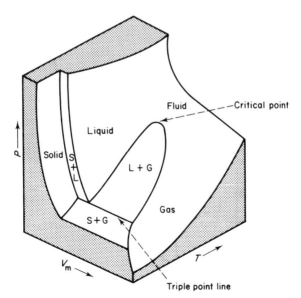

Figure 4.8. The equation of state surface $\phi\,(P, V_{\mathrm{m}}, T)$ for a typical one-component system

90

equation-of-state surface on the (P, T)-plane. As the two-phase regions of the surface have zero slope along the volume axis they appear as boundary lines in the (P, T)-phase diagram. The critical point is at the top of the liquid–gas boundary line so that in Figure 4.8 it is the point of maximum pressure and temperature on the liquid + gas region of the surface.

Figure 3.7 shows typical isotherms of the equation-of-state surface, that is it shows the line of intersection of the (P, V_m, T)-surface with (P, V_m)-planes for different temperatures. Only isotherms above the melting point of the solid were shown in Figure 3.7 but lower temperature curves could have been added which would include the solid region.

We complete this chapter by examining the (P, T)-phase diagrams of two systems, carbon dioxide and sulfur, which are shown in Figures 4.9 and 4.10. The only unusual feature of the CO_2 phase diagram is that the triple point occurs at a pressure greater than 1 atm ($T_t = 217$ K, $P_t = 5.11$ atm). Thus at normal pressures solid CO_2 sublimes and it is this property that gives it its name 'dry ice' and its value as a dry refrigerant. Cylinders of CO_2 at room temperature contain the liquid with a vapour pressure of 67 atm, but when they are discharged a CO_2 snow is produced as a result of the cooling produced by the rapid vaporization of the liquid.

The new feature introduced for sulfur (Figure 4.10) is the existence of two solid phases, rhombic and monoclinic. There are three true triple points

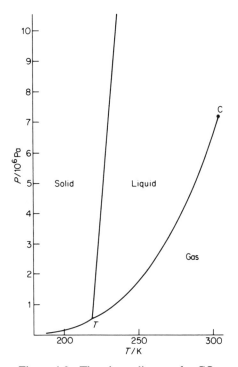

Figure 4.9. The phase diagram for CO_2

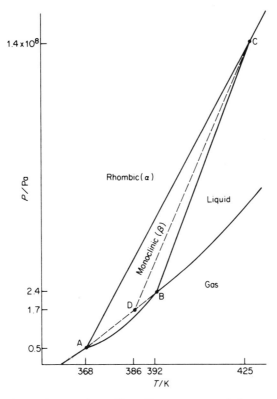

Figure 4.10. The phase diagram for sulfur. The pressure axis is not to scale. A, B and C are true triple points. D is a pseudo-triple point for metastable rhombic, liquid and gas

shown in the figure as A, B and C and one pseudo-triple point which is the conjunction of metastable rhombic, liquid and gas phases. The transition rhombic to monoclinic is sufficiently slow that rhombic sulfur melts before passing to the monoclinic form. Rhombic sulfur can be produced by freezing the liquid above 1.4×10^8 Pa (1400 atm) or by rapidly condensing the vapour onto a cold surface. In practice it is most commonly produced by crystallization from carbon disulfide.

Liquid and gaseous sulfur are known to be mixtures of a number of molecular species. Molecules are known with both rings and chains of sulfur atoms and the species S_2, S_4, S_6 and S_8 are the most common forms. However, the phase diagram is not influenced by this complexity. Whether the liquid consists of a single type of molecule or many types, it is still a single phase because it is uniform in chemical composition.

Chapter 5

Liquid Crystals — The Mesophase

5.1 ORDER IN LIQUID CRYSTALS

The first four chapters of this book have been concerned with the properties of pure liquids. Before turning to the more complex systems of liquid mixtures we discuss a state of matter which is intermediate between that of the crystalline solid and the liquid. This gives us the opportunity of amplifying some of the points made in the early chapters and at the same time we cover a field which in recent years has acquired considerable technological importance.

In 1888 Reinitzer prepared some esters of cholesterol and noted that they had unusual melting properties. Although the crystals had a sharp melting point the melt was opaque but this opacity disappeared suddenly at a higher temperature to give a normal clear liquid.

Both he and a contemporary, Lehmann, established that the opaque state possessed anisotropic properties. This is most clearly shown by viewing thin sections between crossed polarizing plates when patterns are seen due to varying degrees of rotation of the plane of polarization by different parts of the section.

Lehmann used the term 'liquid crystal' for this state of matter to show that it combines the properties of fluidity with that of anisotropy which is a characteristic of some crystalline solids. Despite much criticism over the years this name has remained. Friedel, in particular, stressed that the state was neither crystalline nor strictly liquid, and in 1922 he advocated the term mesophase (*meso* meaning between or intermediate) which is also widely used today.

It should not be thought that substances which exhibit a mesophase are rare. Many hundreds were prepared in the early part of this century and they were widely studied by physical chemists. Indeed, when one under-stands the origins of the phase one is tempted to conclude that most molecules should show a mesophase in some range of temperature and pressure, although there will be far fewer which show a mesophase at atmospheric pressure.

In this chapter we discuss the establishment of the mesophase by an appropriate choice of temperature—so-called thermotropic mesomorphism. It is also possible to establish it by an appropriate choice of solvent. This lyotropic mesomorphism will be considered in a later chapter as it relates to other topics, such as surfaces, colloids and membranes.

A crystal is characterized by long-range order. This is associated with a translational symmetry that not only maintains a regularity in the lattice spacing in three dimensions, but also a regularity in the orientations of the molecules in space. In other words, the crystal possesses long-range translational order of the molecular centres and orientational order of the individual molecules.

In a mesophase some but not all of the long-range order is maintained. For a completely general molecule having three non-equivalent moments of inertia, the orientational order can be maintained along only one axis, with random rotation about that axis, or along all three axes. In addition, the translational order can be maintained in one, two or three dimensions. There are therefore several levels of order between the crystalline state, in which all long-range order is maintained, and the isotropic liquid in which no long-range translational or orientational order is maintained.

From optical studies Friedel was able to distinguish three different types of mesophase, and although further subdivisions are in common use these are still the main categories used today.

(1) The smectic mesophase. This is a viscous state with certain properties similar to those of soaps, from which the name is derived (Greek *smectos*, soap-like).

(2) The nematic mesophase. This is a mobile state which on glass surfaces frequently adopts a characteristic threaded pattern clearly visible between crossed polarizing plates. The name is likewise from the Greek: *nematos*, thread-like.

(3) The cholesteric mesophase. This is, like the nematic mesophase, a mobile state, but it is characterized by bright colours exhibited in reflected light and by a very high optical rotatory power. This phase is predominantly associated with derivatives of cholesterol, but it can be obtained from other chiral substances. To form a cholesteric mesophase it is apparently necessary for the molecule to have a plate-like structure as, for example, cholesterol (Figure 5.1), and to be optically active.

There have been arguments over whether the cholesteric mesophase should be considered as a subclass of the smectic or of the nematic types. Many cholesteric compounds exhibit both cholesteric and smectic phases but none has been found to give both cholesteric and nematic. It appears therefore, that the molecular shapes required for cholesteric and nematic are

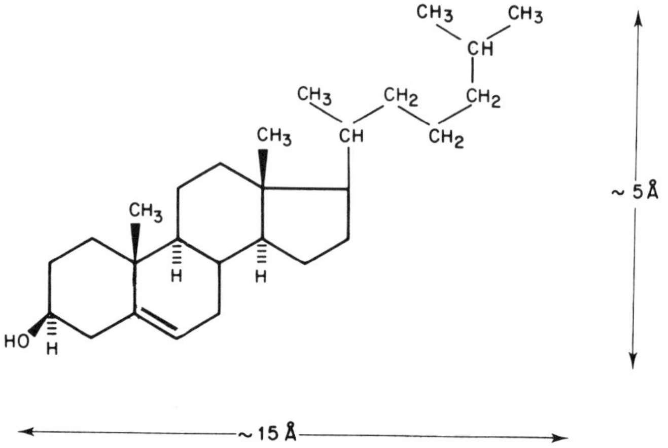

Figure 5.1. The formula of cholesterol

incompatible. However, in fluidity the cholesteric is much closer to the nematic, and the addition of certain optically active compounds to some nematic mesophase will give cholesteric properties. On balance the present view is that the cholesteric mesophase should be considered as a distinct class.

Order in Smectic and Nematic Mesophases

It has already been mentioned that a molecule may exhibit more than one mesophase, and this is particularly true for long molecules which can give smectic and nematic mesophases. On melting crystals of such molecules one obtains first the smectic mesophase, which on further heating turns into the nematic mesophase, and finally at a higher temperature the isotropic liquid. One can therefore conclude that the smectic mesophase has a higher degree of order than the nematic.

Friedel, on the basis of his optical studies and some similarities between the shapes of thin drops of smectic fluid and soap films, suggested that the smectic mesophase possessed a layered structure. This was later confirmed by X-ray studies and the layer thickness was found to be approximately equal to the length of the molecule. In contrast, X-ray studies on the nematic mesophase have shown little evidence of crystallinity, the patterns being similar to those obtained from the isotropic liquid (Section 1.3). One important difference, however, is that X-ray studies carried out in the presence of applied electric or magnetic fields show that molecules in the nematic mesophase are aligned by the field. It can be concluded from this that the nematic mesophase consists of swarms of about 10^5 molecules and within each swarm there is approximate alignment but no layered structure. In other words in each swarm the molecules form an interlocking structure.

The relationship between the molecular arrangements in the crystal, the mesophase and the isotropic liquid is shown schematically in Figure 5.2. However, it must be emphasized that the orientational order in the nematic mesophase occurs only over microscopic dimensions ($\sim 100\,\text{nm}$) in the absence of applied fields.

There are many examples of molecules which exhibit more than one smectic mesophase. For example, the long molecule 4'-n-hexadecyl-3'-nitro-biphenyl-4-carboxylic acid shows three smectic phases with transition temperatures as follows:

$$\text{crystal} \xrightarrow{389.5\,\text{K}} \text{smectic I} \xrightarrow{434\,\text{K}} \text{smectic II} \xrightarrow{466.5\,\text{K}} \text{smectic III} \xrightarrow{471.5\,\text{K}} \text{liquid}$$

Although the smectic mesophases all possess layered structures, the orientation of the molecules within a layer may be different, for example, the molecules may be tilted in some but not in others, and there may be a layer periodicity which is twice that of the simple structure by having head-to-head and tail-to-tail contacts. These are illustrated in Figure 5.3.

Liquid-crystal Displays

The nematic mesophase has in recent years been widely adopted for liquid-crystal display devices (LCDs). They operate through their interaction with natural light either in transmission or reflection, and because they are not themselves primary light providers they consume very little power.

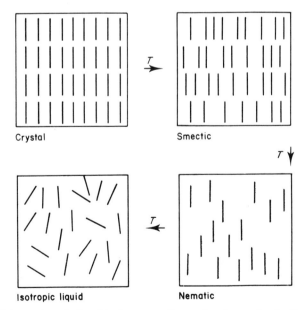

Figure 5.2. Schematic relationship between the molecular order in the smectic and nematic mesophases and in the crystal and isotropic liquid. Perfect alignment in the mesophases should not be inferred

 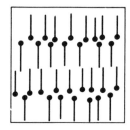

Figure 5.3. Possible alternative smectic structures to that shown in Figure 5.2

One type of display makes use of the fact that the nematic mesophase can be aligned by contact with surfaces, particularly glass surfaces. A small electric field can then be used to disturb this alignment and the change can be seen in reflected light. Another device makes use of the fact that larger fields can induce turbulence in the mesophase which produces enhanced light scattering.

The order in the cholesteric mesophase is shown most clearly by the extremely large optical rotatory power of such systems. For a normal optically active organic liquid rotations of 300° per mm would be considered large but values up to 30,000° per mm have been obtained for some cholesteric mesophases. Such large values show that there must be an ordered packing of the optically active molecules so that there is a cooperative rotation of the electric vector of the light from one molecule to another as it passes through the medium.

A simple model which illustrates the cooperative effect is shown in Figure 5.4. It consists of a set of screwed rods which are packed together so that their screws interlock along an axis perpendicular to the screw axis. This packing generates a helical array of the rods. Molecules similarly packed would produce an exceptionally large optical rotation. Figure 5.4 also shows a schematic representation of the helical structure that is obtained by gradually twisting the axis of alignment of nematic-type layers.

If the pitch of the cholesteric helix, which is the distance along the twist axis for a complete revolution of the molecular axis, is approximately equal to some wavelength in the visible spectrum then light of this wavelength will be reflected strongly by the cholesteric phase. The wavelength for maximum reflection depends on both the pitch of the helix and the angle between the helical axis and the incident ray, the connection being given by an equation similar to that of the Bragg diffraction law.

With suitable compounds the reflected colours can be obtained in any part of the visible spectrum, and in fact the pitch in the cholesteric mesophase can range from 200 nm to 8000 nm. Moreover, the pitch changes with temperature, generally decreasing as the temperature increases, so that the reflected colours act as a sensitive temperature indicator. This has found technical and medical use in thermography, a technique whereby the

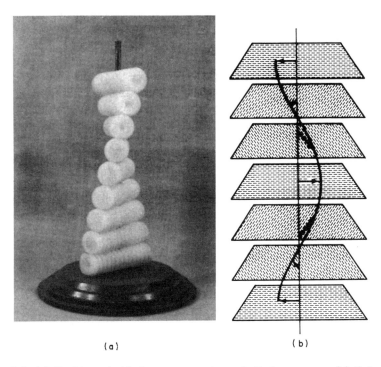

(a)   (b)

Figure 5.4. (a) Packing of chiral screws to give a helical structure. (b) Schematic representation of the helix produced by gradually twisting the axis of alignment of nematic-type layers

temperature of spots on surfaces are shown up by a surface coating of a cholesteric mesophase.

In concluding this section we note that there are crystalline phases in which the translational order of the molecular centres is maintained in three dimensions but in which orientational order is lost. These are frequently found for molecules which are nearly spherical so that little energy is required to allow free rotation or rotational randomization of the molecules in the crystal lattice: tertiary-butanol is one example. These phases are referred to as plastic crystals because their resistance to flow is considerably less than that of their low-temperature normal crystalline form. The entropy of fusion of a plastic crystal is generally low because it possesses a greater entropy than normal crystals by virtue of the rotational randomization of molecules in the lattice.

5.2   MOLECULAR STRUCTURE AND THE MESOPHASE

If one wishes to establish a phase in which long-range translational order of the molecular centres is lost, but long-range orientational order is maintained, then one requires a strong rotational anisotropy in the intermolecular potential. Thus liquid crystals are formed predominantly by molecules that are plate-like or rod-like.

Anisotropy can occur both in the long-range (Section 2.2) and in the short-range (Section 2.3) parts of the potential. One important component to the long-range potential is the electrostatic interaction between molecules, but because liquid crystals can be formed by either polar or non-polar molecules, electrostatic forces cannot be the only factor although they are undoubtedly often important.

The *para*-polyphenyls of formula

$$\left(-\!\!\!\!\!\!-\right)_n$$

are one important class of non-polar molecules which give a mesomorphic phase for $n > 3$. It has been noted that replacing an unsaturated phenyl ring by the saturated group

changes the mesophase transition temperatures considerably. This replacement changes the long-axis polarizability considerably by removing the π-electron delocalization, and hence changing the dispersion energy. However, it makes little difference to the size of the molecule, and we can, therefore, conclude that the anisotropy of the dispersion energy plays an important role in establishing the mesophase. This is supported by the fact that saturated long-chain organic molecules rarely form a mesophase whereas their unsaturated analogues frequently do.

The relationship between the intermolecular potential and the occurrence of a mesophase can be investigated by computer simulation using the methods described in Section 3.5. However, the computational requirements are quite high because the intermolecular potential depends on the relative orientations of the molecular as well as on their distance apart. The first system to be investigated was the orientational order of rigid rods whose centres are fixed at simple cubic lattice sites. This model was introduced by Maier and Saupe[†] to study the isotropic liquid → nematic phase change, but it is in fact a plastic crystal → crystal phase change. The results nevertheless are of interest. Onsager had earlier explored the effects of anisotropy in the hard-core repulsion of rod-like molecules using the traditional methods of statistical mechanics.[‡]

Luckhurst and Romano[§] were the first to make Monte Carlo calculations on needle-like molecules, using an anisotropic attractive potential. They

[†]W. Maier and A. Saupe, *Zeit. Naturf.*, **a15**, 287 (1960).
[‡]L. Onsager, *Ann. N. Y. Acad. Sci.*, **51**, 627 (1949).
[§]G. R. Luckhurst and S. Romano, *Proc. Roy. Soc.*, **A373**, 111 (1980)

started from an isotropic Lennard-Jones potential (2.18) characterized by a well-depth parameter ε and a distance-scaling parameter σ, and they added to this an anisotropic term

$$V_{ij}(R, \theta_{ij}) = -\lambda\varepsilon\left(\frac{\sigma}{R}\right)^6 (3\cos^2\theta_{ij} - 1)/2 \tag{5.1}$$

where θ_{ij} is the angle between the major axes of the molecules. They obtained a nematic phase with short-range translational order for a value of the anisotropy parameter equal to 0.6 at a reduced temperature given by

$$T^* = \frac{kT}{\varepsilon} < 0.9 \tag{5.2}$$

More extensive Monte Carlo calculations were made by Frenkel[†] and co-workers in which the molecules were modelled by hard ellipsoids of various length-to-width ratios (x). Small values of x represent plate-like molecules, and large values needle-like molecules. They obtained a nematic phase for $x \leqslant 0.5$ and $x \geqslant 2.0$ over a range of densities which increased as x departed from these limits. Values of x between 0.5 and 2.0 represent nearly spherical molecules, and there was no density region showing a nematic phase, although a plastic-crystal phase was found. The hard-sphere model is an extreme example of an anisotropic repulsive potential, so we can deduce from both of these studies that nematic phases may result from anisotropy in either the attractive or repulsive parts of the intermolecular potential.

The variation of melting point with chain length in a homologous series is frequently a see-saw function of chain length. This is illustrated in Figure 5.5 by the melting points of the p-n-alkoxybenzoic acids. In contrast, the transition temperatures between mesophases, or between a mesophase and the isotropic liquid, is invariably a smooth function of chain length. In the example shown in Figure 5.5 the nematic-to-liquid transition temperature shows an oscillation from odd to even chain lengths, but the odd and even series lie on separate smooth curves. This phenomenon can be understood as follows.

The melting point of a compound is the temperature at which a three-dimensional ordered structure collapses to a disordered liquid, and the change is substantial. Small differences between molecules can make large changes in the way they pack in the crystal, and hence in the melting point. The change in structure between two mesophases or between mesophase and liquid is, however, relatively small, and as a result the transition temperatures are regular.

This raises the question of whether the phase changes are transitions of first or second order. In the case of p-azoxyphenetole the nematic–liquid transition has been shown to be accompanied by a small discontinuous

---

†D. Frenkel, B. M. Mulder and J. P. McTague, *Mol. Cryst. Liq. Cryst.*, **123**, 119 (1985).

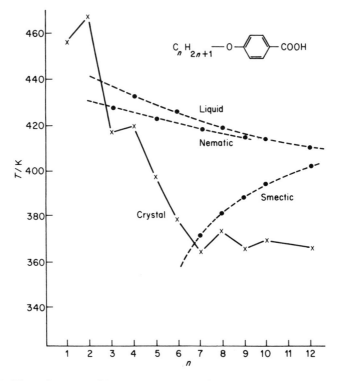

Figure 5.5. The phase-transition temperatures for the p-n-alkoxybenzoic acids (G. W. Gray and B. Jones, *J. Chem. Soc.*, 4179 (1953)). Crosses show the melting points and dots the smectic–nematic and nematic–liquid transition points

change in volume of 0.6% compared with the solid–nematic volume change of 8.4%. Likewise small discontinuities in the enthalpy have been measured—typically only a quarter of the change found on melting. The evidence is, therefore, that transitions between mesophases or between mesophase and liquid are first-order but with small discontinuities in V, H and S.

Liquid crystals can be found in a much higher temperature regime by incorporating suitable rod-like or plate-like components into the structure of a polymer in either or both of two distinct ways. The rod- or plate-like moiety is often called the mesogen or rigid segment in these cases, and it may be incorporated into either the main chain (or backbone) of the polymer molecule or attached to the main chain in side-chains. Schematically, the resulting structures are shown in Figure 5.6. It is important that there should also be flexible elements in the molecules, and the occurrence of liquid-crystal character requires a critical balance between rigidity and flexibility to be maintained. Depending on details of structure, the liquid-crystal phases may be nematic, smectic or cholesteric.

(a)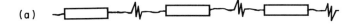

(b)

Figure 5.6. Schematic representation of liquid-crystal polymers of the (a) main-chain and (b) side-chain varieties. The figure depicts: backbone (or main chain)⌇; mesogen ▭; and flexible spacer ⟋⟍ .

Chapter 6

Mixtures of Non-electolytes

6.1 THERMODYNAMIC PROPERTIES OF LIQUID MIXTURES

In this chapter we shall be considering some important properties of liquid mixtures such as their vapour pressure and miscibility. For the moment we confine our attention to liquids which are non-electrolytes, that is, liquids which do not significantly conduct electricity. The behaviour of such liquids closely follows certain ideal laws of mixing, and hence their thermodynamic properties can often be predicted accurately. In contrast, electrolytes, typically solutions of salts in water, show large deviations from these ideal laws, and they will be discussed in Chapters 8 and 9.

In addition to liquid mixtures we shall also consider in this chapter solutions of solid non-electrolytes in liquid non-electrolytes because thermodynamically the two types of system are very similar. The word solution is used to describe a liquid phase containing more than one substance, when one substance which is in excess (the solvent) is treated differently from the others (the solutes).

We start by defining some terms which are measures of the composition of a mixture. The most important quantity, as far as the laws of mixing are concerned, is the mole fraction. The mole fraction of a species A is defined as the ratio of the number of moles of A to the total number of moles of all species present in the mixture. This is written

$$x_A = n_A / \sum_i n_i, \qquad (6.1)$$

where n_i is the number of moles of species i. It follows from definition 6.1 that the sum of the mole fractions of each species in a mixture is equal to unity

$$\sum_i x_i = 1. \qquad (6.2)$$

To make up a solution with specified mole fractions requires each component to be weighed to precalculated amounts, and this can be a tedious operation. There are other measures of composition that are simpler for pratical use. One that is commonly met in making standard solutions for volumetric analysis is the molarity, which is the number of moles of a solute in $1\,dm^3$ of solution. However, for reasons which will become clear later when we discuss changes in volume on mixing, it is not easy to determine mole fractions for a solution of specified molarity.†

A better practical measure of composition is the molality (m) which is the number of moles of a solute in a solution containing $1\,kg$ of solvent. If the molecular weight of the solvent is M then $1\,kg$ of solvent contains $1000/M$ moles, hence the mole fraction of a solute in an m molal solution is

$$x = \frac{m}{m + \dfrac{1000}{M}}. \tag{6.3}$$

For dilute solution $m \ll 1000/M$ and we can use the approximation

$$x \approx \frac{mM}{1000}. \tag{6.4}$$

Miscibility and Immiscibility

If two pure gases are placed in a container but separated by a partition, then when the partition is removed the gases will mix to give a gas which is uniform in composition throughout the container; in other words they give a single phase. Let us now discuss in very general terms the question of whether or not two liquid phases will spontaneously mix.

Ethanol and water will form a single phase whatever their relative proportions; we say that the two liquids are completely miscible. In contrast benzene and water when mixed will form two liquid layers (two phases) unless one of the two components is present in a large excess. Benzene has a very small solubility in water and water a very small solubility in benzene so over most of the mole fraction range benzene and water are essentially immiscible.

If two liquids are miscible in all proportions we conclude that the Gibbs free energy of the mixture is less than the sum of the Gibbs free energies of the separate components. Suppose we make a mixture with n_A moles of component A and n_B moles of component B. Before mixing the total Gibbs free energy is

$$n_A G_{A,m} + n_B G_{B,m}. \tag{6.5}$$

---

†To convert from a volume-based measure (molarity) to a mass-based measure (mole fraction) it is necessary to know the density of the mixture.

We write the free energy after mixing (at constant T and P) as

$$(n_A + n_B)G_m(x), \tag{6.6}$$

where $G_m(x)$ is the molar Gibbs free energy for a mixture with mole fraction of A equal to

$$x = \frac{n_A}{n_A + n_B}. \tag{6.7}$$

We define the Gibbs free energy of mixing as the difference between 6.6 and 6.5 per mole of mixture

$$\begin{aligned}\Delta G_{mix}(x) &= G_m(x) - \left(\frac{n_A}{n_A + n_B}\right)G_{A,M} - \left(\frac{n_B}{n_A + n_B}\right)G_{B,m}, \\ &= G_m(x) - xG_{A,m} - (1 - x)G_{B,m}, \tag{6.8}\end{aligned}$$

$\Delta G_{mix}(x)$ must be negative for all x if the two liquids are miscible in all proportions. In fact we can make a stronger statement than this, as will be seen from the following analysis.

Figure 6.1 shows the typical behaviour of the Gibbs free energy of mixing of two liquids that are completely miscible. Suppose we now take a fraction λ moles ($\lambda < 1$) of a mixture having a mole fraction x_1 of A and $(1 - \lambda)$ of a mixture with mole fraction x_2 of A. Before mixing, the total free energy of mixing is

$$\overline{\Delta G_{mix}}(x) = \lambda\Delta G_{mix}(x_1) + (1 - \lambda)\Delta G_{mix}(x_2) \tag{6.9}$$

which we note is a linear function of λ. Thus $\overline{\Delta G_{mix}}(x)$ is represented by a point on the straight line joining $\Delta G_{mix}(x_1)$ and $\Delta G_{mix}(x_2)$, whose composition, x, is given by

$$\frac{x - x_1}{x_2 - x} = \frac{1 - \lambda}{\lambda}. \tag{6.10}$$

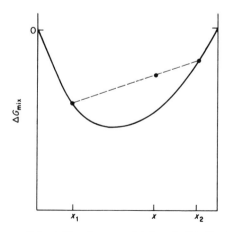

Figure 6.1. ΔG_{mix} for completely miscible liquids

When we mix the two solutions of composition x_1 and x_2 we get a single phase, so we conclude that $\Delta G_{\text{mix}}(x)$ is less than $\overline{\Delta G_{\text{mix}}}(x)$. If it were greater, then the two fractions of compositions x_1 and x_2 would not be miscible.

We contrast the situation of Figure 6.1 with that of Figure 6.2. We see that ΔG_{mix} is negative over the whole range but it is greater than $\overline{\Delta G_{\text{mix}}}(x)$ over the range $x_1 < x < x_2$ as shown. In other words any mixture that has a composition in this range will spontaneously separate into two liquid phases, one of composition x_1 and other of composition x_2. We conclude that for two liquids to be miscible over the whole composition range not only must $\Delta G_{\text{mix}}(x)$ be negative, but it must, in its whole composition range, have a positive curvature with respect to x.

Other thermodynamic mixing functions can be defined in a similar way to ΔG_{mix}. For mixing at constant temperature and pressure the most important quantities are ΔH_{mix} and ΔS_{mix}, which are related to ΔG_{mix} by

$$\Delta G_{\text{mix}}(x) = \Delta H_{\text{mix}}(x) - T\Delta S_{\text{mix}}(x). \qquad (6.11)$$

If ΔH_{mix} is negative and ΔS_{mix} positive the two liquids will be miscible at all temperatures. This conclusion must be qualified by the requirement that the graph of $\Delta G_{\text{mix}}(x)$ against x has positive curvature for the liquids to be miscible in all proportions. If both ΔH_{mix} and ΔS_{mix} are positive the liquids may be immiscible at low temperatures but miscible at high temperatures. The minimum temperature at which the liquids are miscible in all proportions is called the upper consolute temperature or upper critical solution temperature (UCT).

Figure 6.3 shows an example of this behaviour for mixtures of benzene and perfluoro-heptane. At temperatures below 386.5 K there will be two liquid phases if the overall composition of the system lies within the

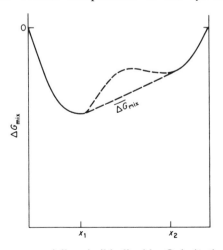

Figure 6.2. ΔG_{mix} for two partially miscible liquids. Only in the regions covered by the solid curves will a single phase occur

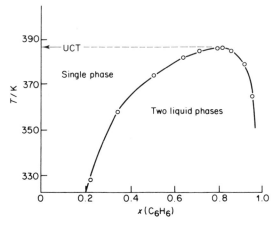

Figure 6.3. The miscibility of liquid benzene and perfluoroheptane (atmospheric pressure). The upper consolute temperature (UCT) is 386.5 K. (J. H. Hildebrand, B. B. Fisher and H. A. Benesi, *J. Amer. Chem. Soc.*, **72**, 4348 (1950))

two-phase region. As T approaches 386.5 K the compositions of these two phases approach one another and at the maximum in the phase-boundary curve ($x(C_6H_6) = 0.83$, $T = 386.5$ K), the compositions become equal. This point therefore has similar thermodynamic characteristics to the critical point in the single-component phase diagram where gas and liquid properties become identical, and it is therefore also called a critical point.

If ΔH_{mix} and ΔS_{mix} are both negative the liquids may be miscible at low temperatures and immiscible at high temperatures. The maximum temperature for which the system shows a single liquid phase is called the lower consolute temperature or lower critical solution temeprature (LCT). Because ΔH_{mix} and ΔS_{mix} are functions of temperature they may change sign as the temperature is varied and there are systems which exhibit both upper and lower consolute temperatures. In some cases LCT < UCT, i.e. at low temperatures the liquids are miscible in all proportions, this is followed by a region of immiscibility and then at high temperatures they are again miscible: nicotine–water is the classic example. In other cases (e.g. methane–n-hexane) the system has LCT > UCT so that there is a middle temperature range in which the components are completely miscible. In principle, all systems showing a LCT should have one higher UCT but only if the pressure is high enough to prevent the liquids boiling before this is reached.

If two liquids of very similar molecular structure are mixed, such as benzene and toluene, they will be miscible in all proportions. ΔG_{mix} will be negative for such systems and we can anticipate that there will be a simple mixing law by which ΔG_{mix} can be predicted. This law is embodied in the concept of an ideal mixture, which we shall approach by first examining the mixing of two gases.

6.2 IDEAL MIXING

Expression 4.25 gives us the chemical potential (molar Gibbs free energy) of a perfect gas as a function of pressure

$$\mu(P) = \mu^{\ominus} \ (P = 1 \ \text{atm}) + RT \ln (P/\text{atm}). \tag{6.12}$$

We will now calculate ΔG_{mix} for two gases at a constant pressure.

Figure 6.4 illustrates the situation; we have n_A moles of gas A at pressure P and n_B moles of B also at pressure P. When mixed there will be $(n_A + n_B)$ moles of gas also at pressure P. By Dalton's law of partial pressures we can say that the partial pressures† of the two gases in the mixture are

$$p_A = \left(\frac{n_A}{n_A + n_B}\right) P, \qquad p_B = \left(\frac{n_B}{n_A + n_B}\right) P. \tag{6.13}$$

Introducing the mole fractions (6.1) these become

$$p_A = x_A P, \qquad p_B = x_B P. \tag{6.14}$$

Before mixing the total Gibbs free energy is

$$G_b = n_A(\mu_A^{\ominus} + RT \ln P) + n_B(\mu_B^{\ominus} + RT \ln P), \tag{6.15}$$

After mixing the total pressure is still P but each gas contributes only a fraction of this, its partial pressure, and hence the Gibbs free energy of the mixture is

$$G_a = n_A(\mu_A^{\ominus} + RT \ln p_A) + n_B(\mu_B^{\ominus} + RT \ln p_B). \tag{6.16}$$

The free energy of mixing is

$$\Delta G_{\text{mix}} = \left(\frac{RT}{n_A + n_B}\right) (n_A \ln (p_A/P) + n_B \ln (p_B/P),$$
$$= RT(x_A \ln x_A + x_B \ln x_B). \tag{6.17}$$

This expression can be generalized in an obvious way for the mixing of any number of non-interacting ideal gases (i) to

$$\Delta G_{\text{mix}} = RT \sum_i x_i \ln x_i. \tag{6.18}$$

An important point to note is that expression 6.18 is independent of

Figure 6.4. Mixing of two perfect gases at the same pressure

†See note on pressure symbols on page 4.

pressure. Thus $(\partial \Delta G_{mix}/\partial P) = 0$ and, from 4.3, it follows that there is no volume change on mixing. This is as expected for ideal gases.

We can obtain an expression for the entropy of mixing from expression 4.4.

$$\Delta S_{mix} = -\left(\frac{\partial \Delta G_{mix}}{\partial T}\right)_P = -R \sum_i x_i \ln x_i. \tag{6.19}$$

From 6.18 and 6.19 we see that

$$\Delta G_{mix} = -T \Delta S_{mix} \tag{6.20}$$

and conclude that $\Delta H_{mix} = 0$. This is also expected; there is no temperature change on mixing ideal gases providing that no chemical reaction occurs.

As x_i has a value between zero and unity, $\ln x_i$ is always negative. It follows that ΔS_{mix} is positive, as expected for a randomizing process, and ΔG_{mix} is negative. Figure 6.5 shows the behaviour of the thermodynamic parameters as a function of mole fraction for two ideal gases. ΔG_{mix} follows the type of curve which, from Figure 6.1, we have already seen to be necessary for two components to be miscible in all proportions.

An ideal system in thermodynamics is one whose parameters follow some precisely defined convenient standard of behaviour. Thus an ideal or perfect gas is one that obeys the perfect-gas laws and the behaviour of real gases can be measured relative to their ideal behaviour. Likewise, ideal mixtures or

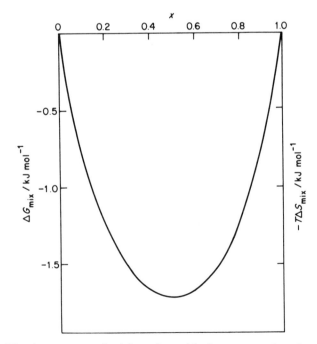

Figure 6.5. The free energy of mixing of two ideal gases as a function of the mole fraction of one component ($T = 300$ K)

ideal solutions are those which follow some standard behaviour, but the appropriate model of ideality will differ according to the type of system under study. In this section of the book we are examining mixtures of non-electrolytes, and for these the ideal system is defined as one whose thermodynamic mixing parameters obey the same equations as an ideal gas: the free energy of mixing should obey 6.18, the entropy of mixing should obey 6.19, and ΔH_{mix} should be zero. We shall later examine other standards of ideality, for example for solutions of electolytes.

Excess Functions

Convenient measures of the non-ideality of a liquid mixture are provided by the excess functions which are defined as the differences between the actual thermodynamic mixing parameters and the corresponding values for the ideal mixture. The most important functions are

$$G^{E} = \Delta G_{\text{mix}} - RT \sum_{i} x_{i} \ln x_{i},$$

$$S^{E} = \Delta S_{\text{mix}} + R \sum_{i} x_{i} \ln x_{i},$$

$$H^{E} = \Delta H_{\text{mix}},$$

$$V^{E} = \Delta V_{\text{mix}}, \tag{6.21}$$

Each of these functions will be nearly zero for mixtures of liquids which are very similar in their chemical composition.

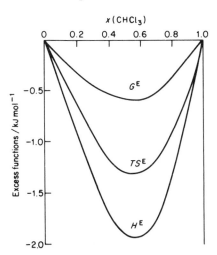

Figure 6.6. The thermodynamic excess mixing functions for acetone and chloroform as functions of the mole fraction of chloroform. (L. A. K. Staveley, W. I. Tupman and K. R. Hart, *Trans. Faraday Soc.*, **51**, 323 (1955))

Figure 6.6 shows the excess functions for mixtures of acetone and chloroform. The negative value of H^E can be attributed to hydrogen bonding between the molecules of the two compounds. ΔS_{mix} will be positive, as mixing is a randomizing process, but S^E is negative because there is more order in the mixture than for an ideal system; again this is due to hydrogen bonding. We shall make further studies of non-ideal mixing later in this chapter.

6.3 PARTIAL MOLAR QUANTITIES

In this section we examine some thermodynamic quantities which measure the change in the extensive properties of a mixture with changing composition. We will take the volume of a liquid mixture as an illustrative example bcause it is relatively easy to see the line of the argument.

A mole of water has a volume of $18 \, \text{cm}^3$, one of ethanol a volume of $58.4 \, \text{cm}^3$. On mixing one mole of each the total volume is $74.5 \, \text{cm}^3$ so that there is a contraction in volume on mixing of $1.9 \, \text{cm}^3$. We shall now define a quantity which can be said to measure the separate volumes of the two components, water and ethanol, in the mixture.

For a two-component mixture (A, B) the total volume is a function of the number of moles of each component. We write this $V(n_A, n_B)$ and think of V as being a surface in three-dimensional space with axes V, n_A, n_B as shown in Figure 6.7. If we make small changes in n_A and n_B the change in volume is given by the complete differential

$$dV(n_A, n_B) = \left(\frac{\partial V}{\partial n_A} \right)_{n_B} dn_A + \left(\frac{\partial V}{\partial n_B} \right)_{n_A} dn_B. \qquad (6.22)$$

To obtain the total volume of a mixture from this expression we have to integrate it from $n_A = n_B = 0$ to the values of n_A and n_B appropriate to the mixture. This can be done easily if we note an important property of V, which is that, provided the composition of the mixture is kept constant (i.e., the ratio n_A/n_B is constant), V is proportional to the number of moles of A (or B). If we return to the water–ethanol example, we know that a mixture of 2 moles of water and 2 moles of ethanol will have a volume $2 \times 74.5 \, \text{cm}^3$. It is this property that gives the surface $V(n_A, n_B)$ the special feature that along a line of constant ratio (n_A/n_B) the slope of V is a constant. This is indicated in Figure 6.7.

To obtain the total volume of a mixture we therefore integrate 6.22 along a line of constant ratio n_A/n_B that will lead to the required mixture. An alternative way to picture this is that we build up the total V by adding small amounts δV, all of which have the same composition $\delta n_A/\delta n_B = n_A/n_B$. By either argument we arrive at the important result that the total volume is given by

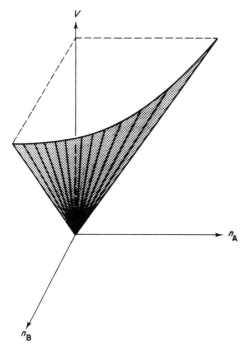

Figure 6.7. A representation of the surface $V(n_A, n_B)$. Lines in the surface of constant composition are shown

$$V = \left(\frac{\partial V}{\partial n_A}\right) n_A + \left(\frac{\partial V}{\partial n_B}\right) n_B. \tag{6.23}$$

The volume of a mixture depends on temperature and pressure as well as composition. As systems are most frequently studied at constant temperature and pressure it is convenient to have a symbol for the change of volume with the concentration of one component, in circumstances where temperature, pressure and the concentrations of all other components are kept constant. This quantity is called the partial molar volume and is given the symbol $\bar{V}$. For the two-component system we have just been discussing we write

$$\bar{V}_A = \left(\frac{\partial V}{\partial n_A}\right)_{T,P,n_B}, \quad \bar{V}_B = \left(\frac{\partial V}{\partial n_B}\right)_{T,P,n_A}. \tag{6.24}$$

Thus for constant (T, P) expression 6.23 can be written

$$V = \bar{V}_A n_A + \bar{V}_B n_B, \tag{6.25}$$

which can be generalized for n components to

$$V = \sum_i \bar{V}_i n_i, \tag{6.26}$$

112

where

$$\bar{V}_i = \left(\frac{\partial V}{\partial n_i}\right)_{T,P,n_j} \tag{6.27}$$

For a pure substance the partial molar volume is equal to the molar volume.

We can summarize the importance of the partial molar volume in two statements. First, from 6.27, it defines the change in volume on changing the amount of one component keeping the amounts of all other components constant. Second, from 6.26, the total volume is equal to the sum of the partial molar volumes of each component multiplied by the number of moles of that component.

The Gibbs–Duhem equation

Figure 6.8 shows the partial molar volumes of ethanol and water as a function of mole fraction in ethanol–water mixtures. We note that the large contractions in volume occurs for the component which is in the mixture only as a trace, the ethanol in almost pure water or the water in almost pure ethanol. Another feature is that there is a maximum in the partial molar

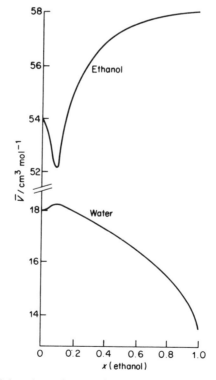

Figure 6.8. The partial molar volumes of an ethanol–water mixture as a function of mole fraction of ethanol

volume of water and a minimum in the partial molar volume of ethanol, and these both occur at the same mole fraction. This result follows from a theorem which we shall now prove.

If we differentiate expression 6.25 with respect to n_A keeping n_B constant we have, at constant T and P:

$$\left(\frac{\partial V}{\partial n_A}\right)_{T,P,n_B} = n_A\left(\frac{\partial \bar{V}_A}{\partial n_A}\right)_{T,P,n_B} + \bar{V}_A + n_B\left(\frac{\partial \bar{V}_B}{\partial n_A}\right)_{T,P,n_B}. \qquad (6.28)$$

However, the left-hand side of this equation is $\bar{V}_A$ so that

$$n_A\left(\frac{\partial \bar{V}_A}{\partial n_A}\right)_{T,P,n_B} + n_B\left(\frac{\partial \bar{V}_B}{\partial n_A}\right)_{T,P,n_B} = 0. \qquad (6.29)$$

Equation 6.29 relates the derivatives of the two partial molar quantities with respect to the number of moles of one component. For more than two components the equation can be generalized to

$$\sum_i n_i\left(\frac{\partial \bar{V}_i}{\partial n_k}\right)_{T,P,n_j} = 0, \qquad (6.30)$$

which is called the Gibbs–Duhem equation for partial molar volumes. It can be seen from 6.29 that if one of the derivatives is zero then the other must be zero also, which explains the coincident maximum and minimum in Figure 6.8.

The equations which we have derived in this section for partial molar volumes can be applied to any of the extensive thermodynamic quantities. For example, the partial molar enthalpy is defined by analogy with 6.24 as

$$\bar{H}_i = \left(\frac{\partial H}{\partial n_i}\right)_{T,P,n_j}, \qquad (6.31)$$

hence the total enthalpy is (cf. 6.26)

$$H = \sum_i \bar{H}_i n_i, \qquad (6.32)$$

and the Gibbs–Duhem equation for enthalpies is (cf. 6.30)

$$\sum_i n_i\left(\frac{\partial \bar{H}_i}{\partial n_k}\right)_{T,P,n_j} = 0. \qquad (6.33)$$

The most important partial molar quantity is the partial molar Gibbs free energy and this is given a special name, the chemical potential, and symbol μ.

$$\left(\frac{\partial G}{\partial n_i}\right)_{T,P,n_j} = \bar{G}_i \equiv \mu_i. \qquad (6.34)$$

The chemical potential of a pure substance is equal to the molar Gibbs free energy, and was defined as such in Section 4.1.

One reason for the impotance of the chemical potential is that for a system containing more than one component,, and comprising two or more phases in equilibrium, the chemical potential of a component must have the same value in every phase. We note that this is an extension of the condition (1.4) for a single component that the molar Gibbs free energy must be equal in the two phases, and the proof of the more general result follows the same line.

6.4 THE VAPOUR PRESSURE OF LIQUID MIXTURES

The vapour pressure of a liquid is defined as the pressure exerted by its vapour under conditions of equilibrium. For a single-component liquid the vapour pressure is a function only of the temperature; the liquid–gas boundary line shown in Figure 1.1 represents this functional relationship. For a liquid consisting of more than one component, the vapour pressure is a function of composition as well as temperature.

The vapour pressure is an extremely important thermodynamic property of a liquid because it provides a measure of its chemical potential. Expression 4.25, or 4.26 if the vapour is not ideal, gives the chemical potential of the vapour as a function of pressure, and if the vapour is in equilibrium with the liquid, the chemical potentials of vapour and liquid are equal. Thus for a pure liquid (species A) we can write (ideal vapour)

$$\mu_A^*(l) = \mu_A(g) = \mu_A^\ominus + RT \ln(p_A^*/atm), \tag{6.35}$$

where the symbol * is used to indicate a pure component. If several components are present in a mixture an expression such as 6.35 applies to each component, as follows

$$\mu_A(l) = \mu_A^\ominus + RT \ln(p_A/atm), \tag{6.36}$$

where p_A is the partial pressure of species A in vapour. Subtracting 6.35 from 6.36 eliminates $\mu^\ominus$, the standard chemical potential of the gas, with the result

$$\mu_A(l) = \mu_A^*(l) + RT \ln(p_A/p_A^*). \tag{6.37}$$

This expression relates the chemical potential of species A to the chemical potential of the pure liquid and to the ratio of the vapour pressures of A in the mixture and in the pure liquid.

$\mu_A(l)$ must be less than $\mu_A^*(l)$ for any stable liquid phase; if it were not, the liquid mixture would be thermodynamically unstable relative to pure liquid A and a mixture with the other components. It follows that p_A must be less than p_A^*, that is the partial pressure of a component in a single-phase mixture is always less than the vapour pressure of the pure component at the same temperature.

Raoult's Law and Henry's Law

We now ask how the ratio (p_A/p_A^*) depends on the mole fraction of component A. The simplest assumption is embodied in Raoult's law

$$p_A/p_A^* = x_A. \tag{6.38}$$

This law implies that the chance of finding a molecule of A in the vapour, relative to the chance for the pure liquid A, is equal to the fraction of molecules A in the mixture. It is evident that this would be a reasonable assumption if we were dealing with ideal mixtures as defined in Section 6.2 (e.g., a mixture of $CHCl_3$ and $CDCl_3$) and in fact Raoult's law can be taken as a definition of ideality, equivalent to that contained in expression 6.18, providing that the vapour is an ideal gas.

Assuming the validity of Raoult's law, expression 6.37 becomes

$$\mu_A(l) = \mu_A^*(l) + RT \ln x_A. \tag{6.39}$$

On mixing n_A moles of A and n_B moles of B, the free energy of mixing per mole of mixture is

$$\Delta G_{mix} = (n_A + n_B)^{-1}[n_A(\mu_A - \mu_A^*) + n_B(\mu_B - \mu_B^*)] = RT(x_A \ln x_A + x_B \ln x_B), \tag{6.40}$$

which is identical to 6.17.

Figure 6.9(a), (b) and (c) shows the vapour pressures of one ideal and two non-ideal mixtures. Note that in the non-ideal case (c), the total vapour pressure is greater than that of either pure component, although we have already noted that the partial vapour pressure cannot be greater than the corresponding pure liquid vapour pressure.

If the vapour pressure of one component obeys Raoult's law then that of the other must also obey this law. This follows from 6.37 and application of the Gibbs–Duhem equation to the chemical potentials. The corresponding expression to 6.30 for two components is

$$\left(\frac{\partial \mu_A}{\partial n_A}\right)n_A + \left(\frac{\partial \mu_B}{\partial n_A}\right)n_B = 0. \tag{6.41}$$

Equation 6.41 can easily be tranformed into a similar expression with mole fractions

$$\left(\frac{\partial \mu_A}{\partial x_A}\right)x_A + \left(\frac{\partial \mu_B}{\partial x_A}\right)x_B = 0. \tag{6.42}$$

If we now relate the chemical potential to vapour pressure by 6.36 we obtain

$$\left(\frac{\partial \ln p_A}{\partial x_A}\right)x_A + \left(\frac{\partial \ln p_B}{\partial x_A}\right)x_B = 0, \tag{6.43}$$

116

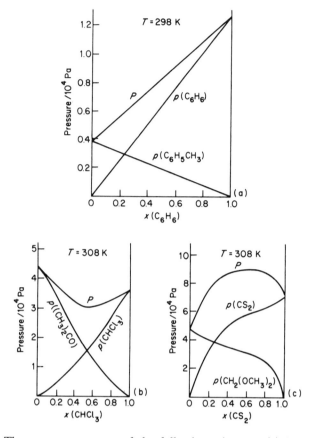

Figure 6.9. The vapour pressures of the following mixtures: (a) benzene–toluene; (b) acetone–choloroform; (c) carbon disulfide–dimethoxymethane

or

$$\frac{x_A}{p_A}\left(\frac{\partial p_A}{\partial x_A}\right) + \frac{x_B}{p_B}\left(\frac{\partial p_B}{\partial x_A}\right) = 0. \tag{6.44}$$

Because $x_A + x_B = 1$ we obtain the result

$$\frac{x_A}{p_A}\left(\frac{\partial p_A}{\partial x_A}\right) = \frac{x_B}{p_B}\left(\frac{\partial p_B}{\partial x_B}\right) \tag{6.45}$$

Equations 6.43–6.45 are known as the Duhem–Margules equations; they establish a relationship between the gradients of the vapour pressure–mole fraction curves of the two components. In particular, if $(\partial p_A/\partial x_A)$ is constant then the left-hand side of 6.45 is equal to unity and $(\partial p_B/\partial x_B)$ must also be a constant.

Even for non-ideal mixtures Raoult's law is obeyed for the excess component in a very dilute solution. This is indicated by the partial pressure line becoming asymptotic to the Raoult's law line as $x \to 1$. For dilute solutions the vapour pressure of the component of *low* concentration is also linearly related to mole fraction, but the proportionality constant does not have the Raoult's law value. This result is known as Henry's law and is expressed mathematically as

$$p_A = K_A x_A, \tag{6.46}$$

when $x_A \to 0$. K_A is known as the Henry's law constant. The physical basis of Henry's law is that for dilute solutions every solute molecule is completely surrounded by solvent molecules and providing this condition is maintained the vapour pressure of the solute will be proportional to the mole fraction of solute. Deviations from Henry's law will occur when concentrations are reached such that solute molecules have other solute molecules in their vicinity.

Henry's law has special significance for the solubility of gases in liquids. Consider, for example, the solubility of hydrogen (component A) in benzene as a function of the applied pressure P. If P is much greater than the vapour pressure of benzene we can take $P \approx p_A$ whence 6.46 takes the form

$$x_A = P/K_A, \tag{6.47}$$

and if x_A is small we can, from 6.4, express the solubility in molality units

$$m_A = \frac{1000}{M_B K_A} P. \tag{6.48}$$

In our example M_B would be the molecular weight of benzene ($78 \, \text{g mol}^{-1}$) and K_A the Henry's law constant for hydrogen in benzene, which is $3.62 \times 10^3 \, \text{atm}$ at 298 K. Thus at a pressure of 10 atm the solubility of hydrogen in benzene is $3.5 \times 10^{-2} \, \text{mol kg}^{-1}$.

Figure 6.10 shows the solubility of some common gases in water as a function of applied pressure. The Henry's law constants are given by the tangents to the curves at the origin; it can be seen that Henry's law applies quite accurately up to gas pressures of 100 atm or greater. At very high pressures a correction should be made for non-ideality of the gas and to do this P in 6.47 is replaced by fugacity (f). It has been found empirically that $\log(f/x)$ is a linear function of P, but with the pressure-dependence being very small.

Henry's law coefficients increase with increasing temperature so that gases become less soluble in liquids as the temperature is raised. A widely observed relationship is $\log K_A \propto T^{-1}$.

Deviations from the ideal mixing laws can often be explained quantitatively in terms of the interactions between molecules in the pure liquids and in the mixture. For example, the negative deviation from Raoult's law observed for acetone–chloroform mixtures (Figure 6.9(b)) is attributed to hydrogen bonding between these two species in the mixture. This is

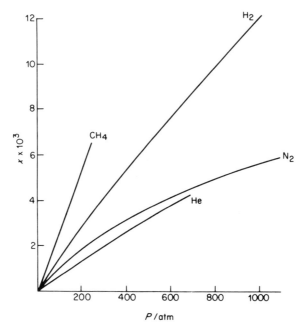

Figure 6.10. The solubility of gases in water as a function of applied pressure ($T = 298$ K)

supported by evidence from the infra-red spectrum of such mixtures, particularly the increase in intensity of the C—H stretching band of chloroform which occurs on addition of acetone. There are also changes in the carbonyl stretching band of the acetone, and these changes have been used to show that a complex with one acetone and two chloroform molecules is present in the mixture.†

Positive deviations from Raoult's law are more common than negative deviations, and occur particularly for mixtures of water with a less polar solvent. The reason for this is that water is a strongly associated liquid and a second component will tend to break up this association with a correspondingly unfavourable change in the enthalpy.

A class of solution which can be thought of as intermediate between ideal and non-ideal was defined by Hildebrand and called a regular solution. For such solutions the entropy of mixing obeys the ideal mixing law (6.19), but the enthalpy of mixing is not ideal ($\Delta H_{mix} \neq 0$). The main interest in this class of solution is that it allows for an elegant statistical mechanical treatment of the free energy of mixing.‡ However, the number of solutions

---

†K. B. Whetsel and R. E. Kagarise, *Specttochim. Acta*, **18**, 329 (1962).
‡See for example G. S. Rushbrooke, *Introduction to Statistical Mechanics*, O.U.P., Oxford (1949).

which are exactly regular is very small: some solutions of I_2 in organic solvents provided the impetus for the concept. A full discussion has been given by Hildebrand and Scott.†

6.5 THE EFFECT OF SOLUTE ON THE BOILING AND FREEZING POINTS OF A LIQUID: COLLIGATIVE PROPERTIES

In this section we shall prove that an involatile solute will lower the freezing point and raise the boiling point of a solvent, and we shall find the values of actual changes as functions of solute concentration on the assumption that the solution is ideal. To calculate the freezing point depression we have to make the additional assumption that the solute does not dissolve in the solid solvent.

The qualitative result is shown very simply. If the solute is involatile and does not dissolve in the solid solvent then it will only affect the chemical potential of the solvent in its liquid phase. Moreover, the solute must lower the chemical potential of the liquid solvent (if not it would not be soluble) so that the situation is summarized in Figure 6.11. Lowering the chemical potential of the liquid from $\mu^*(l)$ to $\mu(l)$ lowers T_m to T'_m and raises T_b to T'_b.

An important application of this result is that the melting point is an important criterion of purity of a compound because impurities lower the melting point of a substance from its value when pure.

Over a small temperature range the decrease in the liquid solvent chemical potential arising from the added solute can be considered to be a constant, hence the temperature-dependence of $\mu(l)$ and $\mu^*(l)$ will be represented by parallel lines. From the expanded version of the gas–liquid crossing point in Figure 6.11 it can be seen that, providing the changes bought about by the solute are small,

$$\mu^*(l) - \mu(l) = AC - BC = CD(\tan \theta_1 - \tan \theta_2)$$

$$= \Delta T_b \left[\frac{-d\mu(g)}{dT} + \frac{d\mu^*(l)}{dT} \right] = -\Delta T_b \frac{d}{dT}(\mu(g) - \mu^*(l)), \quad (6.49)$$

But, from 4.5,

$$\mu^*(l) - \mu(l) = \Delta T_b \Delta S_{vap} = \frac{\Delta T_b \Delta H_{vap}}{T_b}, \quad (6.50)$$

the last step being due to the fact that $\Delta G_{vap} = 0$.‡

From the expanded version of the solid–liquid crossing point we likewise have

---

†J. H. Hildebrand and R. L. Scott, *Regular Solutions*, Prentice Hall, Englewood Cliffs, New Jersey (1962).
‡ΔG_{vap} etc. are molar quantities.

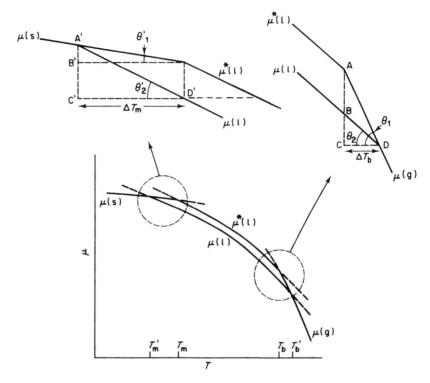

Figure 6.11. Effect of an involatile solute on the melting and boiling points of the solvent

$$\mu^*(l) - \mu(l) = A'C' - A'B' = C'D'(\tan \theta_2' - \tan \theta_1')$$

$$= \Delta T_m \left[-\frac{d\mu^*(l)}{dt} + \frac{d\mu(s)}{dT} \right] = -\Delta T_m \frac{d}{dT}(\mu^*(l) - \mu(s)) \quad (6.51)$$

$$= \Delta T_m \Delta S_{melt} = \frac{\Delta T_m \Delta H_{melt}}{T_m}. \quad (6.52)$$

For an ideal solution we can write (6.39)

$$\mu^*(l) - \mu(l) = -RT \ln x_1, \quad (6.53)$$

where x_i is the mole fraction of the solvent. If x is the mole fraction of solute, then

$$\ln x_1 = \ln(1 - x) = -x - \frac{x^2}{2} \ldots, \quad (6.54)$$

and in dilute solution we need to take only the first term in the expansion and write

$$\mu^*(l) - \mu(l) = RTx. \quad (6.55)$$

This expression can now be introduced into 6.50 and 6.53, taking T as T_b

or T_m as appropriate. We therefore obtain the following expressions for the elevation of boiling point and depression of melting point due to the added solute

$$\Delta T_b = \left(\frac{RT_b^2}{\Delta H_{vap}} \right) x, \tag{6.56}$$

$$\Delta T_m = \left(\frac{RT_m^2}{\Delta H_{melt}} \right) x. \tag{6.57}$$

It can be seen from expressions 6.56 and 6.57 that the changes in boiling and melting point depend on specific properties of the solvent (T_b, T_m, ΔH_{vap}, ΔH_{melt}) but only on the mole fraction of the solute; no other feature of the solute is relevant. Solvent properties of this type are called colligative properties.† One other colligative property, the osmotic pressure, will be described in a later chapter.

The most important practical use of a colligative property is for the determination of molecular weights in solution. For dilute solutions we can replace the mole fraction by molality according to expression 6.4. Expression 6.57 then becomes

$$\Delta T_m = K_t m, \tag{6.58}$$

where

$$K_t = \frac{MRT_m^2}{1000 \Delta H_{melt}}, \tag{6.59}$$

is called the cryoscopic constant of the solvent. The corresponding constant for boiling point elevation is called the ebullioscopic constant. Measuring ΔT_m for a solvent of known cryscopic constant determines m, and if the mass of solute in the solution is known then the molecular weight of the solute can be determined.‡

The factor M, which is the molecular weight of the solvent, appears in the definition of the cryoscopic constant (6.59) hence it is advantageous to use solvents of high molecular weight for the determination of colligative properties, particularly if the solute is sparingly soluble. For example, benzene has a cryoscopic constant of $5.1 \text{ K kg mol}^{-1}$, and that of camphor is 40 K kg mol^{-1}; water has a very low cryoscopic constant, $1.86 \text{ K kg mol}^{-1}$.

Ideal Solubility

It is convenient in this section to consider also the question of the ideal solubility of a solid in a solvent, because the mathematics of the answer is

---

†From the Latin *colligatus*, bound together. The word can be thought of as having the connotation that these are collective properties of all solutes.
‡For electolytes in ionizing solvents each ion will contribute separately to a colligative property so that measuring the depression of freezing point (say) will determine the degree of ionization or, perhaps, the nature of the ionic equilibria if these are uncertain.

similar to that we have just described. A solid will dissolve in a solvent until its chemical potential in the solution equals the chemical potential of the pure solid. Starting again with expression 6.39 and taking species A to be the solute, we have for a saturated solution (assuming ideality)

$$\mu_A^*(s) = \mu_A(l) = \mu_A^*(l) + RT \ln x_A \tag{6.60}$$

whence (dropping the A for brevity)

$$\ln x = \frac{\mu^*(s) - \mu^*(l)}{RT} = \frac{-\Delta G_{melt}(T)}{RT}. \tag{6.61}$$

We know that the Gibbs free energy of melting is zero at the melting point (T_m) of the solute, but it is not zero at temperature T below T_m. If we add a term of zero value to 6.61 as follows

$$\ln x = -\left\{ \frac{\Delta G_{melt}(T)}{RT} - \frac{\Delta G_{melt}(T_m)}{RT_m} \right\} \tag{6.62}$$

then, assuming ΔH_{melt} and ΔS_{melt} do not change appreciably over the temperature range $T \to T_m$, we simplify 6.62 to

$$\ln x = \frac{\Delta H_{melt}}{R} \left[\frac{1}{T_m} - \frac{1}{T} \right]. \tag{6.63}$$

Close to the melting point $(T \approx T_m)$ this can be exponentiated to

$$x \approx \exp\left[\frac{\Delta H_{melt}(T - T_m)}{RT_m^2} \right] \tag{6.64}$$

hence the solubility, expressed as mole fraction of solute, increases

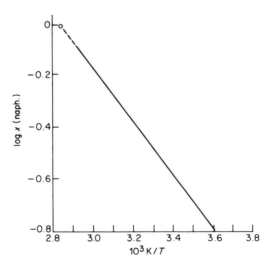

Figure 6.12 The solubility of naphthalene in benzene as a function of temperature. The extrapolation is to the melting temperature

exponentially with temperature, becoming unity at $T = T_m$. Figure 6.12 shows the solubility of naphthalene in benzene as a function of temperature, and confirms the functional form of expression 6.63.

Expressed in mole fractions, ideal solubility is independent of the solvent but, in molality units, ideal solubility is greater for solvents of small molecular weight than for those of large molecular weight.

6.6 THERMODYNAMIC MEASURES OF NON-IDEAL MIXING

Earlier in this chapter we introduced two definitions of ideal liquid mixtures and showed them to be equivalent;† the thermodynamic mixing functions should obey equations 6.18 and 6.19 and the vapour pressure should obey Raoult's law, 6.38. Deviations from ideality can therefore be expressed in terms of the mixing functions or in terms of the vapour pressure, the link between the two being made through expression 6.37

$$\mu_A(l) = \mu_A^*(l) + RT \ln (p_A/p_A^*). \tag{6.65}$$

For brevity we will drop the suffix indicating a specific component and also take the liquid phase to be understood so that 6.65 is written

$$\mu = \mu^* + RT \ln (p/p^*), \tag{6.66}$$

the star indicating the pure liquid. Expression 6.66 is exact providing the vapour is behaving as an ideal gas. If it is not then the vapour pressures must be replaced by fugacities (f) as defined by 4.26 and 4.27.

For ideal mixtures the ratio of the vapour pressures can be replaced by the mole fraction x, to give 6.39. For non-ideal mixing it is convenient to retain the form of expression 6.39 by introducing a dimensionless quantity, a, called the activity, or more precisely the relative activity, which appears in place of x as follows:

$$\mu = \mu^* + RT \ln a. \tag{6.67}$$

Providing that the vapour is behaving as an ideal gas, we have

$$a = p/p^*. \tag{6.68}$$

If we wish to record the chemical potential of a component in a non-ideal mixture we can either have a table of μ values as a function of composition or, what is common practice, a table of the activity as a function of composition, with μ^* being the chemical potential of the reference state. We refer to μ^* as being the chemical potential of the standard state. In this case the standard state is the pure liquid and $\mu = \mu^*$ when $a = x = 1$. We have

---

†Assuming that the vapour is ideal.

already noted that the solvent in a dilute solution obeys Raoult's law as infinite dilution is approached so that

$$a \to x \quad \text{as} \quad x \to 1. \tag{6.69}$$

An important consequence of the mathematical form of equation 6.67 is that μ approaches minus infinity as a approaches zero, because a approaches zero as x approaches zero. In other words the chemical potential of the *solute* in a very dilute solution has a large negative value and this is another reason why a is a more convenient quantity than μ as a measure of non-ideality, because a is bounded by the limits zero and one.

The relationship between a and x can further be illuminated by introducing a quantity called the activity coefficient, γ, by the definition (cf. 4.28)

$$a = \gamma x. \tag{6.70}$$

As a and x are both dimensionless, γ is also dimensionless. For very dilute solutions we have for the solvent, from 6.59, the limit $\gamma \to 1$. However, for the solute in very dilute solutions there is no unique limit to γ because

$$\gamma = a/x \tag{6.71}$$

is indeterminate when both x and a approach zero. We can make progress by remembering that in dilute solutions the vapour pressure of the solute obeys Henry's law (6.46). Thus introducing 6.68 and 6.46 into 6.71 we have

$$\gamma \to K/p^* \quad \text{as} \quad x \to 0, \tag{6.72}$$

where K is the Henry's law constant and p^* is the vapour pressure of the pure solute.

Because the solute in a dilute solution is very far from the $x = 1$ limit it is more convenient to take the reference or standard state of the solute as a solution obeying Henry's law. This is particularly true for a solute which is not a liquid (when pure) at the temperature being studied. Introducing Henry's law into 6.66 we have

$$\mu = \mu^* + RT \ln\left(\frac{Kx}{p^*}\right),$$

$$= \mu^* + RT \ln\left(\frac{K}{p^*}\right) + RT \ln x,$$

$$= \mu^\ominus + RT \ln x, \tag{6.73}$$

where

$$\mu^\ominus = \mu^* + RT \ln\left(\frac{K}{p^*}\right). \tag{6.74}$$

Expression 6.73 is now the basis for a suitable approach to the chemical potential of the solute in dilute solutions; following our earlier practice we replace x in 6.73 by a and obtain

$$\mu = \mu^{\ominus} + RT \ln a = \mu^{\ominus} + RT \ln \gamma x. \tag{6.75}$$

This has a similar form to 6.67, but the standard state has the chemical potential $\mu^{\ominus}$ and not μ^*. We also note that the Henry's law limit implies that

$$a \to x \quad \text{or} \quad \gamma \to 1 \quad \text{as} \quad x \to 0 \tag{6.76}$$

which should be contrasted with 6.69, the limit which is applicable when the pure solvent is taken as standard state.†

It is probably more common to express solute concentrations in molalities than in mole fractions, and when this is done the definitions of activity and activity coefficient are slightly different from those given here. We return to this point in Section 9.4.

It cannot be stressed too strongly that values of the activity or activity coefficient have meaning only when the relevant standard state is specified, and failure to note this will make meaningless a comparison between different systems on the basis of their a or γ values. For example, one cannot even compare a given solute in different solvents without recognizing that the standard state of the solute is solvent-dependent (because the Henry's law constant is solvent-dependent).

What system actually has the chemical potential $\mu^{\ominus}$? The answer is no real system. We note that $\mu - \mu^{\ominus}$ when $a = 1$ but this is not then a system obeying Henry's law. The standard state is a hypothetical state for which $a = 1$ *and* Henry's law is obeyed. The situation is comparable to that described in Section 4.2 for non-ideal gases: the standard state of a real gas is an extrapolated ideal gas at 1 atm.

We shall make a further generalization of the concept of activity in Chapter 9 when we consider non-ideal behaviour for electrolyte solutions, and we hope that these repeated discussions will help to clarify a very tricky concept in thermodynamics.

Determination of Activity

To determine the activity of the solvent in a solution is generally straightforward if the solute is involatile. One measures the vapour pressure and applies expression 6.68. If the solute is volatile it is necessary to

---

†Another measure of non-ideality that is frequently used is the osmotic coefficient ϕ defined by

$$\mu = \mu^{\ominus} + \phi RT \ln x. \tag{6.76a}$$

Comparison with 6.75 shows that ϕ is related to the activity coefficient by

$$\ln \gamma = (\phi - 1) \ln x. \tag{6.76b}$$

The name stems from the fact that ϕ for the solvent is equal to the ratio of the actual osmotic pressure of a solution to its ideal value. Osmotic pressure, a colligative property, will be discussed in Chapter 11.

determine the partial pressure of the solvent. Only for the most accurate work would it be necessary to make a correction for the non-ideal behaviour of the vapour. To determine the activity of a non-volatile solute is much more difficult. Electrolyte activities can be obtained from e.m.f. measurements as will be explained in Chapter 10. For non-electrolytes it is usually necessary to use an indirect method which entails finding the activity of the *solvent* over a range of solute concentrations and then integrating the Gibbs–Duhem equation.

Introducing activities in place of chemical potentials in 6.42, using 6.67, we have for a two-component mixture

$$\left(\frac{\partial \ln a_A}{\partial x_B}\right) x_A + \left(\frac{\partial \ln a_B}{\partial x_B}\right) x_B = 0. \tag{6.77}$$

If A is a solvent for which a_A has been determined by vapour pressure measurement (say) as a function of x_B, the mole fraction of solute, then integrating 6.77 with respect to x_B from some specified value x_B^0 to the concentration of interest we have

$$\int_{x_B^0}^{x_B} \frac{x_A}{x_B} \left(\frac{\partial \ln A}{\partial x_B}\right) dx_B + \left| \ln a_B \right|_{x_B^0}^{x_B} = 0. \tag{6.78}$$

Because $x_A = 1 - x_B$ this can be simplified to

$$\ln\left(a_B(x_B)/a_B(x_B^0)\right) = -\int_{x_B^0}^{x_B} \left(\frac{1 - x_B}{x_B}\right)\left(\frac{\partial \ln a_A}{\partial x_B}\right) dx_B. \tag{6.79}$$

Hence if we know the activity of the solute at a concentration x_B^0 and the activity of the solvent between x_B^0 and x_B, we can determine $a_B(x_B)$.

There are two situations in which the activity of the solute is known. First, in a saturated solution the chemical potential of the solute is equal to that of solid and μ_B(solid) can be obtained by standard thermochemical methods (most can be found in thermodynamic tables). Second, in dilute solutions the solute would satisfy Henry's law and hence expression 6.73 can be applied. Thus if $\mu_B^{\ominus}$ has been determined, μ_B or a_B can be obtained in more concentrated solutions.

Because the value of a colligative property is (ideally) proportional to the mole fraction of solute it might be thought that measurements of such properties would give directly the solute activity. That is, one could simply replace x by a in an equation such as 6.57. This is unfortunately not the case. The solute mole fraction only appears in these equations by assumptions which imply that the solutions are both dilute and ideal (e.g. terminating the expansion 6.54 at the first term). Colligative properties are in fact direct measures of the activity of the *solvent* and, as for the vapour pressure, it is necessary to use the Gibbs–Duhem equation to deduce the activity of the solute.

The non-ideal form of the melting point equation 6.52 is

$$\ln a_A = -\frac{\Delta T_m \Delta H_{melt}}{RT_m^2}, \tag{6.80}$$

where ΔT_m is a function of solute concentration. it follows that

$$\frac{d \ln a_A}{dx_B} = \frac{-\Delta H_{melt}}{RT_m^2} \left(\frac{d\Delta T_m}{dx_B}\right), \tag{6.81}$$

and this can be inserted into 6.79 and the integration performed,

$$\ln (a_B(x_B)/a_B(x_B^0)) = \frac{\Delta H_{melt}}{RT_m^2} \int \left(\frac{1 - x_B}{x_B}\right) \left(\frac{d\Delta T_m}{dx_B}\right) dx_B. \tag{6.82}$$

Mixing and Vapour Pressure

We have seen that the ideal solubility of a solid in a liquid can be expresssed in terms of the mole fraction of the substance in the saturated solution through equation 6.63. This result is based on the principle that the chemical potential of substance A (the solute) must have the same value in the solid state and in the saturated solution. It is also true that substance A must have the same value of chemical potential in the vapour phase as in the saturated solution, and parallel arguments apply to substance B (the solvent), hence we may write, from equations 6.35–6.37,

$$\mu_A^*(s) = \mu_A^*(l) + RT \ln x_A = \mu_A + RT \ln (p_A/p_A^*)$$

and

$$\mu_B(l) = \mu_B^*(l) + RT \ln x_B = \mu_B + RT \ln (p_B/p_B^*).$$

From these equations it is apparent that the limiting solubility of a solid in a liquid is a function of the partial pressures of the two components in the vapour over the saturated solution and to the vapour pressures of the separate pure components. The way in which solubility follows vapour pressure is particularly important in understanding the function of super-critical fluids (see below), which have remarkable solvent properties in relation both to the amount of solute they can accommodate and the degree of selectivity between solutes.

Cohesive Energy Density and the Solubility Parameter

It is noted elsewhere in this book (page 189) that broad generalizations can be made to define which solutes dissolve in (mix with) which solvents but that it is impossible to formulate precise rules which permit solubility or insolubility to be predicted with confidence. The classic foundations of studies in this field were laid by Hildebrand, who assigned the principal role to a property designated internal pressure. Since mixing is inevitably accompanied by an increase in entropy, the positive or negative character of the concomitant free energy change depends upon the enthalpy factor, and much effort has been devoted to relating the enthalpy change on mixing to measurable properties of the separate components. Probably the most successful attempt was that of Hildebrand and Scott, who deduced the equation below.

$$\Delta H_m = V_m[(\Delta U_1/V_1)^{1/2} - (\Delta U_2/V_2)^{1/2}]^2 \phi_1 \phi_2 \tag{6.83}$$

Here, ΔH_m is the enthalpy of mixing, V_m is the volume of the mixture, and ΔU, V, and ϕ are, respectively, the internal energy change on vaporization, the molar volume, and the volume fraction of a component. $(\Delta U_1/V_1)$ is thus the energy of vaporization per unit volume of component 1, and it was this quantity which was originally termed the internal pressure; later it acquired the name cohesive energy density, still used today. It is a measure of the amount of energy necessary to overcome all the intermolecular forces holding the molecules together in unit volume of the substance, given that the internal energy of the vapour is negligible. If equation 6.83 is rearranged in the following manner,

$$\Delta H_m/(V_m \phi_1 \phi_2) = [(\Delta U_1/V_1)^{1/2} - (\Delta U_2/V_2)^{1/2}]^2 \tag{6.84}$$

it is seen that the enthalpy of mixing per unit volume at a given composition is equal to the square of the difference between the square roots of the cohesive energy densisties of the two components. This fact prompted Hildebrand and Scott to make the substitution

$$\delta = (\Delta U/V)^{1/2} \tag{6.85}$$

and to call the new quantity δ, the solubility parameter. Since

$$\Delta H_m/(V_m \phi_1 \phi_2) = (\delta_1 - \delta_2)^2 \tag{6.86}$$

it follows that enthalpic considerations will not prevent mixing from occurring if $(\delta_1 - \delta_2)$ is relatively small, and that mixing should certainly occur if $(\delta_1 - \delta_2)$ is zero. Thus, two substances are likely to mix freely if they have similar values of the solubility parameter, a thermodynamic justification for the generalisation that 'like dissolves like'.

Values of solubility parameter can be determined from the enthalpy of vaporization and in other ways. The values for non-polar substances are relatively low and for polar materials relatively high; a typical median value for a normal liquid is $19 \, J^{1/2} \, cm^{-3/2}$. Many systems conform to the general rule that mixing occurs when the solubility parameters are close in value but allowance must be made for particular interactions, such as polarity or hydrogen bonding, which require a correction factor to be added to the value of the solubility parameter as deduced from the enthalpy of vaporization.

6.7 SOLUBILITY AND THE SUPERCRITICAL STATE

In Chapter 1 we discussed the supercritical state briefly, and drew attention to some of its salient features. It is appropriate at this point to say a little more about the use of supercritical fluids as solvents, i.e. in two-component systems. Extraction by supercritical fluids, first employed commercially in 1978, is used industially to remove caffeine from coffee, to remove nicotine from tobacco, and to extract pharmaceutically-useful substances from natural materials. Supercritical fluids can also be used as solvents for chemical reactions, thus many organic solvents and oxygen will dissolve to form a single-phase system in which very efficient oxidation takes place; such processes are being studied as ways of coping with the large-scale disposal of domestic waste.

In general, supercritical fluids (Xe and CO_2 are common examples) have many advantages as solvents. They are often extremely good solvents, their solvent power can be adapted by variation of temperature or pressure, and they are usually pharmacologically inert. Moreover, they can exercise great discrimination between solutes; thus CO_2 exercises a selectivity of almost 100 with respect to cholesterol and ergosterol, despite the similarity in their structures (shown in Figure 6.13). Both these substances have extremely low vapour pressures, of the order 10^{-10} bar, nevertheless careful measurement of these vapour pressures reveals that they are exactly parallel to their solubilities in supercritical CO_2. In order to understand the nature of solubility in supercritical fluids, it is necessary to examine the basic physical chemistry of such systems a little more closely.

We have seen that the liquid–gas boundary line in the (T, P) phase diagram for a single-component system terminates abruptly at the critical point, the coordinates of which can be written as (T_c, P_c). At any point where $T > T_c$ and/or $P > P_c$, there is no distinction between gas and liquid and, if either T or P is reduced (the other being maintained constant), the two-phase gas–liquid boundary line is not crossed. Figure 3.7 likewise shows that, when $T > T_c$, there is no region in the (P, V) isotherms in which gas and liquid co-exist. A single-component system in this condition is known as a supercritical fluid.

The first point to note is that solubility is frequently a complicated

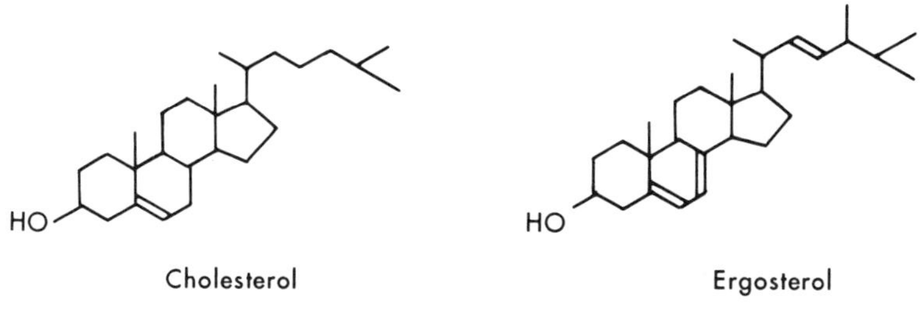

Figure 6.13. Structures of cholesterol and ergosterol.

function of external pressure at constant temperature. Starting from the solvent as a low-pressure gas (in which the solubility of any solute will be low), solubility will increase with increasing pressure up to a maximum, beyond which it will decrease again. By contrast, when the temperature is increased at constant pressure, solubility increases with temperature according to the 'ideal' law $\ln x \propto T^{-1}$ (equation 6.63), *provided that the pressure is adjusted continuously to maintain the solvent at constant density*. This is illustrated in Figure 6.14, which shows how the solubility of fluoranthrene varies with T^{-1} in supercritical CO_2 at various densities. This figure also shows the variation of the vapour pressure of fluoranthrene with reciprocal temperature; it conforms to the normal empirical law (equation 4.18); we shall comment on the significance of this fact later in relation to equation 6.87.

A careful examination of the available experimental data shows that the solubility of a solute in a supercritical fluid is a very sensitive function of the solvent density ρ. Figure 3.7 demonstrates that $dP/dV_m = 0$ at the critical point; as the density is proportional to V_m^{-1}, it follows that, under the same conditions, $dP/d\rho$ will also be equal to zero, i.e. that $d\rho/dP = \infty$. Thus, the density is extremely sensitive to pressure at, or just above, the critical point, and the solubility of a solvent in a supercritical fluid will consequently vary rapidly with pressure in this region. Thus, for example, by raising the

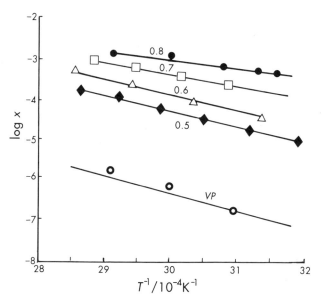

Figure 6.14. The solubility of fluoranthrene in supercritical CO_2 as a function of reciprocal absolute temperature at various CO_2 densities (indicated in $g\,ml^{-1}$). The vapour pressure of fluoranthrene is also shown in the same temperature regime. (R. Burk and P. Kruus, *Can. J. Chem. Eng.*, **70**, 403 (1992).)

pressure to 400 bar, the solubility of squalane in CO_2 is increased by *ten orders of magnitude*, from about 10^{-11} to 10^{-1} $g\,cm^{-3}$. The strong dependence of density (and therefore solvent capacity) on temperature and pressure thus provides an opportunity to exercise remarkable control over the power of the solvent to dissolve a particular substance or to discriminate between solutes.

Similar reasoning can be applied through the concept of the solubility parameter and its relationship to density, and hence pressure. The solubility parameter of a supercritical fluid, and hence its potential for compatibility with other components in a mixture, can be varied continuously over a wide range, from values characteristic of gases to values characteristic of liquids, by adjustment of the pressure and temperature. Thus, for example, δ for CO_2 is almost zero in the gas phase but has a value in the upper part of the normal range for the liquid, and it can be adjusted to any desired intermediate value at will in the supercritical region. More generally, the solubility in supercritical fluids of virtually non-volatile substances is found to be as much as seven orders of magnitude higher than would be expected on the basis of the mixing of ideal gases. Raising the pressure applied to a supercritical fluid increases both its density and its solubility parameter, and thus enhances its tendency to dissolve polar solutes. This concept underlies the use of supercritical fluids as solvents in extraction processes and as mobile phases in chromatographic separations. The solubility parameter of a supercritical fluid can be related to the critical pressure and the reduced densities of its gas and liquid states. Thus, if the solubility parameter of the solute is known, it is possible to predict which supercritical fluid could be brought to conditions in which it would have a solubility parameter close to the same value and hence would act as a solvent. The use of pressure to obtain the necessary conditions is particularly attractive because of the extreme sensitivity of density to pressure in the supercritical region.

An empirical expression, with some basis in thermodynamics, which relates density and the temperature-dependance of solubility in supercritical fluids has been proposed by Chrastil[†], and is displayed here as equation 6.87,

$$c = d^k \exp(a/T + b) \qquad (6.87)$$

where c is the solubility in $g\,l^{-1}$, d is the density in $g\,l^{-1}$, and k, a, b are constants characteristic of both solute and solvent. k is referred to as the association number and it corresponds to the average number of solvent molecules associated with a solute molecule. For supercritical CO_2, k has a wide range of values, from 1.58 for H_2O to 12.1 for cholesterol, for example. In general, k increases with the size of the solute molecule, a fact which has been correlated with clustering of solvent molecules around the solute, a supposition supported by spectroscopic measurements[†]. In the

---

[†]J. Chrastil, *J. Phys. Chem.*, **86**, 3016 (1982).

Table 6.1. Some physical properties of typical liquids, gases and supercritical fluids

Property	Liquid	Supercritical fluid	Gas
Diffusivity/cm^2 s^{-1}	10^{-5}	10^{-3}	10^{-1}
Viscosity/g cm^{-1} s^{-1}	10^{-2}	10^{-4}	10^{-4}
Density/g cm^{-3}	1	10^{-1}	10^{-3}

highly compressible region close to the critical point, solute molecules can induce a very large increase in the density of the solvent (at constant pressure), an effect which is formally associated with a large *negative* partial molar volume of the solute.‡

The efficient use of solvents in extraction and separation processes depends not only on the solubility parameter but also on transport properties such as diffusion constant and viscosity; some idea of the extent of the variation in these properties that can be utilized in the supercritical range is apparent from Table 6.1. Typically, in the supercritical state, a fluid will have the viscosity of a gas combined with a diffusivity intermediate between that of a gas and a liquid. It should therfore be possible to make separation columns with very small theoretical-plate heights (see Chapter 7).

†S. Kim and K. P. Johnston, *AIChE Journal*, **33**, 1603 (1987).
‡C. A. Eckert, D. H. Ziger, K. P. Johnston and S. Kim, *J. Phys. Chem.*, **90**, 2738 (1986).

Chapter 7

Phase Diagrams for Multicomponent Systems

7.1 A GENERAL STATEMENT OF THE PHASE RULE

In Section 1.2 we discussed the concept of degree of freedom for a single chemical substance and its relevance to the equilibria involving gas, liquid and solid phases. In this chapter we extend this study to multicomponent systems with particular emphasis on three techniques of great practical importance, distillation, crystallization and chromatography.†

We first explain how the term component is used in a discussion of the phase rule. The number of components is defined as the *minimum* number of independent chemical substances which, if allowed to mix and come to equilibrium, will form the system in question. If there are no chemical reactions between the chemical species of which the mixture is composed the number of components is just equal to the number of chemical substances. If there are reactions, then we have to ask whether there are any equilibria established on the time-scale of the experiment which inter-relate the actual species present.

For example, a mixture of hydrogen, oxygen and water is normally a three-component system because the equilibrium

$$2H_2 + O_2 \rightleftharpoons 2H_2O,$$

is established extremely slowly. In the presence of a catalyst which would establish this equilibrium rapidly it would be a two-component system because the equation

$$K = [H_2O]^2/[H_2]^2[O_2], \tag{7.1}$$

provides a linking condition that allows only two of the three concentrations

---

†In this chapter we consider only the equilibrium aspects of phase change. Kinetic aspects, such as nucleation, will be dealt with in Chapter 12.

to be varied independently. By similar reasoning an aqueous solution of sodium chloride is a two-component system although the actual species present are Na^+, Cl^- and H_2O. The reason for this is that the number of anions must be equal to the number of cations to preserve neutrality; in other words their concentrations cannot independently be varied. We emphasize that, in the terminology of the phase rule, the number of components is often fewer than the number of chemically distinguishable species which are present.

As explained in Section 1.2, the number of degrees of freedom for a system having just one component is two (which can be considered to be temperature and pressure) minus the number of independent linking conditions of equal chemical potential between the phases. If there are c components, the composition of each phase is specified by $c - 1$ mole fractions (one mole fraction is fixed by specifying all others (6.2)) so that the composition of p phases is described by $p(c - 1)$ mole fractions. However, these cannot be independently varied because of the condition that the chemical potential of any one of the components must be equal in all phases. Thus there are $c(p - 1)$ independent restricting equalities between the chemical potentials. The total number of degrees of freedom for a system of c components and p phases is therefore

$$f = 2 + p(c - 1) - c(p - 1),$$

which simplifies to

$$f = 2 + c - p \tag{7.2}$$

This is Gibbs' phase rule.

We will now consider the equilibria between phases in a two-component system. We remind the reader that the number of degrees of freedom is the number of variables that can be changed without changing the number or type of the phases which are in equilibrium. An equivalent statement is that f is the dimensions of the variable space within which a certain equilibrium set of phases can exist.

For two components we have $f = 4 - p$ and the following situations are of interest.

(1) $p = 1$, $f = 3$. A single phase can exist under changes in P, T and mole fraction, x of a component.

(2) $p = 2$, $f = 2$. Only two of x, P, and T can be varied independently without changing the two phases which are in equilibrium. For example, suppose vapour and a single liquid phase are in equilibrium, then for a specified composition of the mixture (x) the vapour pressure is a function of temperature. That is, there is a curve in (P, T)-space along which the two phases can exist for a specified x, just as there is a (P, T)-curve for these two phases for a pure substance (see Figure 1.1).

(3) $p = 3$, $f = 1$. A typical situation might be vapour and two liquid phases (immiscible liquids) or vapour, liquid and solid. If we fix one

variable, say take $P = 1$ atm, then the other two are invariant. Three phases could only exist for a specified pressure at fixed values of temperature and composition.

(4) $p = 4$, $f = 0$. This situation is neither important nor easily realized. For example, there is, in principle, a point in (x, P, T)-space where the immiscible liquids benzene and water are in equilibrium with vapour and one solid phase. It is analogous to the triple point in the one-component case.

Later in this chapter we will discuss some of the features of three-component systems which have an additional degree of freedom in each of the situations described above. For the moment we turn to an analysis of the equilibrium between liquid and vapour in a two-component system. This analysis will be based upon the phase rule, and the rules governing the vapour pressure of liquid mixtures which were described in Section 6.4.

7.2. VAPOUR–LIQUID EQUILIBRIUM WITH TWO COMPONENTS

We have seen in case (2) of the previous section that a two-component system having a single liquid phase in equilibrium with vapour has two degrees of freedom. That means that in a three-dimensional axis system (x, P, T) this particular equilibrium can exist on a two-dimensional surface. For example, the pressure will be a function of T and x.

Three-dimensional figures are not easy to construct and analyse, and it is usual practice to describe the behaviour of two-component systems by two-dimensional sections of the (x, P, T)-diagram, either (x, P) or (x, T); we have already used (x, P)-sections in Figure 6.9. Figure 7.1(a) shows the vapour pressure of an ideal two-component liquid at a fixed temperature as a function of the composition of the liquid phase expressed as the mole fraction of one component. The straight line plot is a manifestation of Raoult's law. To think of this as a phase diagram we consider a liquid of arbitrary composition x_1 under a pressure P_1. As this pressure is greater than the vapour pressure of a liquid of this composition (P_1'), the system is 100% liquid. If we now reduce the pressure, the first bubbles, indicating boiling, will appear when the pressure reaches P_1'.

For a single-component liquid the pressure would remain at P_1' until all the liquid was evaporated. This is not the case for a mixture because the vapour will be richer than the liquid in the more volatile component, which in this case is B, because B has a higher vapour pressure than A. If we preferentially boil off more B than A the liquid will become progressively richer in A and its composition will change in the direction indicated in the diagram.

Figure 7.1(b) shows the vapour pressure line as a function of the mole

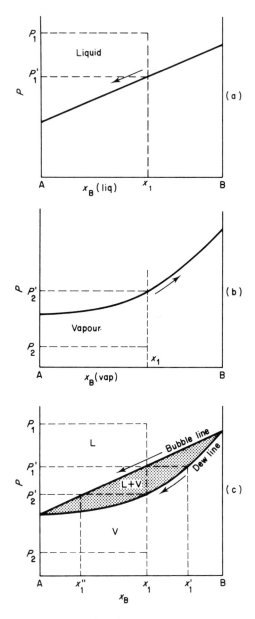

Figure 7.1. Representations of the (P, x) section of a liquid–vapour phase diagram

fraction in the vapour phase. It is concave upwards for the reason given above: the vapour is richer than the liquid in the more volatile component $(x_B(\text{vap}) > x_B(\text{liq}))$ at a given pressure. Suppose we take the vapour composition x_1 at a pressure P_2. As P_2 is less than the vapour pressure we will have 100% vapour. On increasing the pressure the first drops of liquid will appear when the pressure is P_2' (the dew point or condensation point). The first liquid that condenses out at this pressure is richer in the less volatile component (A), and as this condensation proceeds the vapour becomes richer in B, as shown.

It is convenient to combine Figures 7.1(a) and 7.1(b) in such a way that x can represent the mole fractions in both the liquid and the vapour phases and this is shown in Figure 7.1(c). It is important to understand that, when liquid and vapour are in equilibrium, a horizontal line drawn through the (x, P) point corresponding to the overall composition of the system intersects the bubble line at the value of x for the liquid phase and the condensation line at the value of x for the vapour phase. A line of this type, identifying the compositions of phases in equilibrium, is called a tie line. In other words the shaded region is a two-phase region where the compositions of the two phases are given by the left-hand and right-hand boundaries.

If we now start with liquid at the point (x_1, P_1) and reduce the pressure, we will reach the bubble line at P_1' and the first vapour will have composition x_1'. On reducing the pressure the compositions of liquid and vapour will move as shown by the arrows and when all the liquid has evaporated the vapour must also have composition x_1. Thus the last drops of liquid will have composition x_1'' as shown. Starting with vapour at (x_1, P_2) and increasing the pressure the reverse route will be followed; the first drop of liquid that condenses has composition x_1'' and the last amount of vapour has composition x_1'.

Suppose we have a system whose overall composition and pressure is represented by a point in the two-phase region (x, P) as shown in Figure 7.2. We know that the compositions of the liquid and vapour phases are given by x'' and x' respectively, but we can go further and deduce the relative proportions of liquid and vapour that will be in equilibrium under these conditions.

Let n be the total number of moles of mixture and let n_1 and n_v be the number of moles in the liquid and vapour phases, respectively. The total number of moles of B is

$$nx = (n_1 + n_v)x. \tag{7.3}$$

However, summing up the amount of B in the liquid and vapour phases we have

$$nx = n_1 x'' + n_v x', \tag{7.4}$$

and on subtracting 7.3 from 7.4 we have

$$n_1(x'' - x) + n_v(x' - x) = 0, \tag{7.5}$$

where

$$\frac{n_v}{n_1} = \frac{x - x''}{x' - x}. \tag{7.6}$$

138

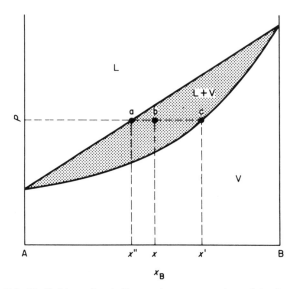

Figure 7.2. Definition of a tie line and representation of the lever rule

Expression 7.6 can be summarized by what is known as the lever rule. The relative amounts of the phases are in inverse proportions of the distances of the respective phase boundary lines from the point (x, P) of overall composition. Consider the line (abc) in Figure 7.2 to be a lever with (b) the fulcrum. If weights are hung at (a) and (c) then for a balance the weight at (a) multiplied by the distance (b − a) would have to equal the weight at (c) multiplied by the distance (c − b). If 'weight at a' is replaced by 'moles of liquid' and 'weight at c' by 'moles of vapour' then the above statement is in accord with expression 7.6.

Distillation

A similar discussion can be applied to any two-component two-phase system (e.g., solid–liquid) and also to (x, T)-sections. We will now examine such an (x, T)-section for the purpose of analysing the technique of distillation. Figure 7.3 shows a typical (x, T)-section of the phase diagram appropriate to the A–B mixture represented in Figure 7.2. At a fixed pressure, B is more volatile than A and has the lower boiling point.

The two-phase region has boundary lines which represent the composition of the vapour and liquid. The bubble line gives the temperature at which the liquid starts to boil. The dew line gives the temperature at which the vapour starts to condense. The bubble line gives the liquid composition, and the dew line the vapour composition of any mixture whose overall composition and temperature lie in the two-phase region.

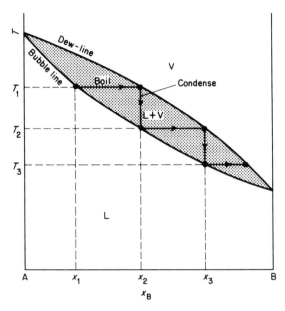

Figure 7.3. The (x, T)-section of a two-component, liquid plus vapour, phase diagram

Let us know consider raising the temperature of a liquid of composition x_1. At T_1 it will start to boil and the first vapour that comes from it will have a composition x_2. If this vapour is condensed and the resulting liquid of this new composition is heated, it will start boiling at T_2, and the first vapour that comes from it will have composition x_3. It can be seen that this process of repeated boiling and condensation leads to a distillate which is richer in the more volatile component, and in the limit of an infinite number of steps would lead to pure B, but a negligible amount of B.

Simple distillation is inefficient for the separation of liquids with close boiling points. Young,[†] for example, describes the distillation of a benzene–toluene mixture in which the distillate was first collected in ten fractions or cuts which were then combined and redistilled fourteen times. The process took thirty hours to obtain appreciable amounts of reasonably pure benzene and toluene. Such a separation can be achieved much more efficiently by the process of fractional distillation in which the compositions of the initial fractions are made as distinct as possible. To achieve this one requires the partial condensation of the vapour over the still and the maximum equilibration between the rising vapour and the returning liquid.

The apparatus for fractional distillation consists of an insulated column

---

[†]S. Young, *Distillation Principles and Processes*, Macmillan, London (1922).

140

which has a loose packing of some inert material, or other device for achieving a large gas–liquid interface, under nearly adiabatic conditions.† This column is topped by a condenser, and the fractions are taken from a side-arm at the top of the column, below the condensing head. The efficiency of the column is defined by the number of effective equilibrium vaporizations and condensations of the type illustrated in Figure 7.3 that are produced within the column: this is called the number of theoretical plates. Columns typically have one theoretical plate for each 1–5 cm of height.

Azeotropes

Figures 7.1 and 7.3 are typical (x, P)- and (x, T)-diagrams for an ideal two-component liquid mixture. For non-ideal mixtures the bubble line in the (x, P)-diagram will be curved (i.e. the system will not obey Raoult's law) and the other phase boundary lines will show greater curvature.

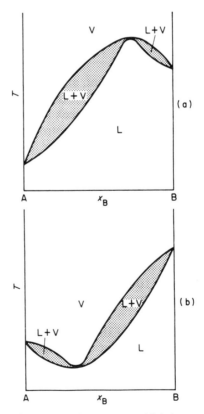

Figure 7.4. (x, T)-diagrams for systems which form an azeotrope

†F. E. Williams, *Techniques of Organic Chemistry*, Eds. E. S. Perry and A. Weissberger **4**, 299, Wiley, London and New York (1965).

Figure 7.4 shows two commonly occurring (x, T)-diagrams for systems with substantial deviation from ideality, but which are still miscible in the liquid phase. Note that in each figure the boundary lines show a coincident turning point so that the two-phase region is divided into two separate parts. The turning points in the phase boundary lines must be coincident to make physical sense. For example, a situation represented by Figure 7.5 would imply that between T_1 and T_2 there would be two immiscible vapours, but we know that vapours are always miscible. If the boundary lines did not touch in Figure 7.4(b) this would imply that there were two liquid phases hence another two-phase region would have to be added to the diagram.

The boiling of a liquid mixture whose composition coincides with the turning point in the phase boundary lines leads to a vapour with the same composition as that of the liquid. In other words, the liquid mixture is behaving as if it were a single component, and removal of the distillate does not lead to a change in the composition of the liquid. Liquid mixtures with this property are called azeotropes (from the Greek for 'boiling without change'). Figure 7.4(a) shows a maximum-boiling azeotrope; chloroform–acetone is such a system and water–HCl is another. Figure 7.4(b) shows a minimum-boiling azeotrope; ethanol–water is an example.

An important point to note is that the azeotropic composition is a function of the pressure, showing that an azeotrope is not a molecular complex with a specific molecular composition.

Figure 7.6 illustrates the route taken in the fractional distillation of a liquid mixture having a minimum-boiling azeotrope. It can be seen that whatever the initial composition of the liquid the final distillate will be the azeotrope, and the residue will be one of the components. Thus aqueous

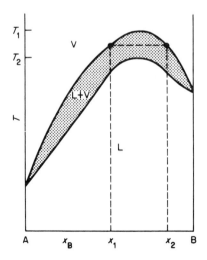

Figure 7.5. An unrealistic (x, T)-diagram which would imply that vapours of composition x_1 and x_2 would be immiscible

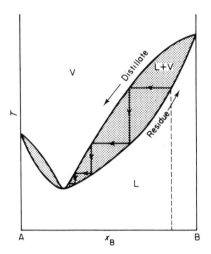

Figure 7.6. The process of fractional distillation of a liquid mixture having a minimum-boiling azeotrope. A path towards the azeotrope is followed for compositions on both sides of the azeotrope composition

ethanol cannot be dehydrated directly by fractional distillation; the result will be the azeotrope which has a composition with 10 mole per cent water. If the ethanol is dried beyond this by other means then a final distillation will give a *residue* of absolute ethanol.

Fractional distillation of systems exhibiting a maximum-boiling azeotrope as in Figure 7.4(a) will produce one of the components as distillate and the residue will be the azeotrope.

We have seen that azeotropes are associated with maxima or minima in the (x, T)-phase boundary lines. Such turning points are associated with deviations from ideality, but as should be clear from Figure 7.4 they are also more likely if the two components have rather similar boiling points. Alternatively, if we think of the (x, P)-diagram (Figure 7.1), an azeotrope is likely at any temperature at which the vapour pressures of the two components are approximately equal.

Partially Miscible Liquids

The liquid–vapour phase diagram for two components which are only partially miscible in the liquid state can have various forms. If the liquids are miscible at the boiling point but immiscible at some lower temperature, the phase diagram will be as in Figure 7.3 or 7.4, but with a two-liquid phase region at some lower temperature. If the two liquids are only partially miscible at the boiling point then a phase diagram such as Figure 7.7 is the most common.

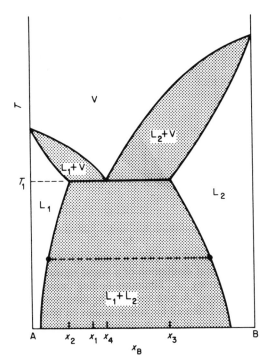

Figure 7.7. The most common (T, x)-diagram for two liquids which are partially miscible at their boiling points (e.g. water–isobutanol). The dashed line in $L_1 + L_2$ is a typical tie line

There is one point in Figure 7.7 where three phases, vapour and two liquids, are in equilibrium; this is at the cusp in the vapour composition boundary line with $x_B = x_4$. In accordance with the phase rule there will be a line of such points in the full (P, T, x)-diagram. This special point in Figure 7.7 is analogous to the azeotropic point in Figure 7.4(b) but there are two important differences. First, the vapour-composition boundary line has a cusp at this point and not a smooth turning point, and second, the azeotropic liquid consists of two phases whose relative amounts are determined by the appropriate tie line in the liquid plus liquid region.

A mixture of composition x_1 will boil at T_1 and at this temperature the liquid consists of two phases with composition x_2 and x_3. These two liquid phases will have definite vapour pressures, the sum of which will be the total pressure (one atmosphere, if the experiment is carried out in the open laboratory). Thus the boiling point will remain at T_1 (which is lower than the boiling point of either pure component), and the composition of the vapour will remain at the azeotropic composition x_4 until one of the two liquids has disappeared. Because in this example $x_4 - x_2$ is less than $x_3 - x_4$, the vapour will have a composition closer to that of L_1 than to that of L_2, hence L_1 will be the first liquid phase to disappear. When this has happened the composition of the remaining liquid phase L_2 will have become richer in B.

If the two liquids are almost completely immiscible the regions L_1 and L_2 shown in Figure 7.7 will be very narrow. The mixture will boil when the total vapour pressure equals the external pressure P and this will be at a temperature lower than the boiling point of either pure component. The composition of the vapour will be given by

$$x_A = \frac{n_A}{n_A + n_B} = \frac{p_A}{p_A + p_B} = \frac{p_A}{P}, \tag{7.7}$$

where p_A and p_B are the partial pressures of the two components at the boiling point.

This analysis provides the basis of the technique known as steam distillation (bubbling steam through a relatively involatile liquid) which allows heat-sensitive materials to be distilled at a lower temperature than their normal boiling point. The technique is widely used for industrial separations, but there are usually preferable alternative separation techniques available for small-scale work (e.g. distillation under reduced pressure).

7.3 LIQUID–SOLID EQUILIBRIUM WITH TWO COMPONENTS

The general form of the vapour–liquid two-component phase diagrams which have been described in the last section apply also to liquid–solid equilibrium. There are, however, a number of important differences. First, liquids and solids have very low compressibilities so that the melting point is a very slowly varying function of pressure. Thus (x, P)-diagrams of the type shown in Figure 7.1 are not very useful for the liquid–solid phase equilibrium because the boundary lines are very steep: if the temperature is fixed at the melting point of one component there is unlikely to be any physically accessible pressure at which the other component melts at the same temperature. Liquid–solid phase diagrams are, therefore, always considered as (x, T)-sections for a fixed pressure.

A second point is that equilibrium in the solid phase is only attained slowly, in contrast to the fast equilibria of the liquid or vapour phases. Consider, for example, Figure 7.8 which is analogous to Figure 7.3. If liquid of composition x_1 is cooled the first solid will appear at temperature T_1, and this will have the composition x_2. As this solid is richer in component A (the higher melting solid) the liquid composition will move to the right as indicated by the arrow. On further cooling more solid will appear and its composition will also move to the right. The important point to note is that continuous cooling produces a solid which is homogeneous in the sense that

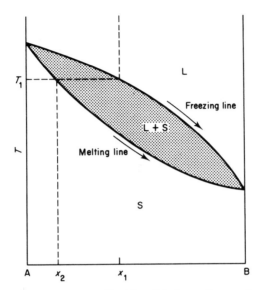

Figure 7.8. A simple two-component liquid–solid phase diagram for which there is a single solid phase (e.g. Ag–Au)

it does not consist of aggregations but it is not of uniform composition. The last trace of solid produced may be almost pure B.

Solidification of a multicomponent liquid is therefore not exactly equivalent to condensation of a multicomponent vapour within physically realizable time scales and this difference is used in the important technique for purifying solids known as zone refining. If a rod of impure solid is passed slowly through a local heater the band of melt will be richer than the solid in any lower melting impurities. These impurities will be swept through with the melt to the end of the rod. Repeated passage through the heater will produce a progressively purer solid.

A third difference between liquid–solid and gas–liquid two-component equilibria is that most pairs of solid components are not miscible. Thus Figure 7.8 does not represent a typical situation, and only occurs if A and B have very similar crystal structures. The solid phases of such systems are often called solid solutions. A more typical liquid–solid phase diagram is analogous to Figure 7.7 for partially miscible liquids. However, the single-phase regions analogous to L_1 and L_2 of Figure 7.6 can be so narrow that they are often omitted altogether. Figure 7.9 shows a typical phase diagram for this situation.

Eutectics

The minimum on the L–S boundary indicates a mixture with low melting point and one that melts without change in composition. This is called an

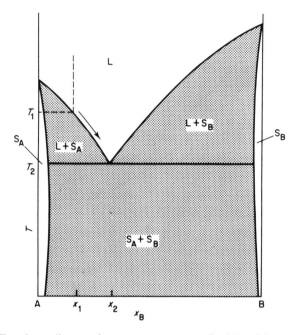

Figure 7.9. The phase diagram for a two-component liquid–solid system with almost immiscible solids (e.g. Ag–Cu)

eutectic (Greek for low melting) and it has a similar role in liquid–solid phase changes to a minimum-boiling azeotrope in gas–liquid phase changes. There are many important eutectics amongst metal alloys. The most common is tinman's solder, a mixture of tin and lead in the approximate ratio 2:1 which melts at 456 K.

Although the eutectic mixture has a definite composition and sharp melting point it is not a single chemical compound. The composition is not exactly represented by a simple integer ratio of the components and separate microcrystals of the two components can be seen under a microscope. In other words it is an intimate mixture of the two components and not a chemical compound.

Cooling Curves

The boundary lines of liquid–solid phase diagrams can be mapped out by carefully following the cooling curves of the liquid. For a single component the ideal form of the cooling curve is shown in Figure 7.10(a). The rate of cooling of the liquid is proportional to the difference between the temperature of the liquid and that of the surroundings. When the melting point T_m is reached the heat of solidification will maintain the system at this temperature until all the solid has formed; after this the solid will cool. If the

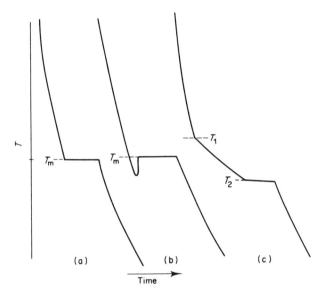

Figure 7.10. Cooling curves. (a) Ideal curve for a single component. (b) Single component showing supercooling. (c) A two-component mixture with a eutectic temperature T_2

rate of cooling is too rapid then the temperature of the liquid may fall below the melting point before the first solid appears. This is called supercooling and it will give a cooling curve as in Figure 7.10(b).

If the liquid of composition x_1 in Figure 7.9 is cooled the first solid (almost pure A) will appear at temperature T_1 (subject to possible supercooling). The liquid will become richer in component B and the equilibrium temperature will fall as indicated. When the liquid composition reaches the eutectic point x_2 its temperature will be T_2 and both solids will separate. The temperature will stay at T_2 until all the liquid has solidified. An ideal cooling curve appropriate to this route is shown in Figure 7.10(c).

Complex Formation

A final complication which we touch on in this chapter is that in the solid phase the two components may form a complex with definite composition (i.e. a chemical species whose composition is independent of temperature and pressure). One example of this is shown in Figure 7.11. The simplest way of analysing such a diagram is to consider it as two diagrams like Figure 7.9 side-by-side.

Figure 7.11 shows the phase diagram for a system which forms a solid compound of formula AB. This compound is stable up to its melting point. Whether AB exists in the liquid state or not is irrelevant as far as the phase diagram is concerned because the liquid would have only two components in

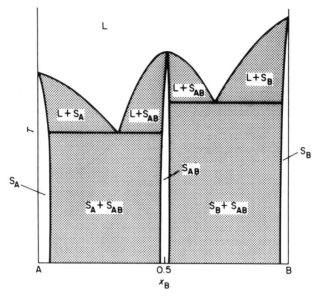

Figure 7.11. Phase diagram for two components which form a solid compound with composition AB (e.g. benzophenone–diphenylamine)

either case, the amount of AB being dependent on the amounts of A and B originally taken. The single-solid phase regions may again be very narrow and are often omitted. However, without them one is likely to be misled about the nature of the two-phase shaded areas, particularly the central hump. A tie line here connects liquid and solid AB and not two liquid phases.

There are more complicated liquid–solid phase diagrams that we will not discuss in detail. For example one may have immiscible liquids and solids, or compounds formed which become unstable before the melting point is reached. A discussion of these can be found in more detailed texts.†

7.4 THREE-COMPONENT LIQUIDS

From the phase-rule expression 7.2 we see that the number of degrees of freedom of a three-component system is

$$f = 5 - p \tag{7.8}$$

There are clearly many features of such a multidimensional system that

---

†F. Thomas and I. Pal, *Phase Equilibria Spatial Diagrams*, Iliffe Books, London (1970); A. Findlay, *The Phase Rule and its Application*, Dover, London (1951).

could be discussed but we will look at only one situation which is the equilibrium between liquid phases at constant temperature and pressure. With (P, T) constant there will be $3 - p$ remaining degrees of freedom. A single phase can exist in a two-dimensional variable space and two phases can exist along a line.

The most convenient method of plotting three-component phase diagrams is to use triangular graph paper which puts each of the three components on the same footing. Figure 7.12 shows this method of plotting the composition of any mixture. Each of the three vertices is identified with one particular component so that a vertex represents 100% of that component and the opposite line 0% of that component. The fraction of component A can therefore be indicated by lines parallel to BC.

Any mixture with fractional composition a, b and c for the three components A, B and C respectively, where $a + b + c = 1$ (these can be mole fractions, volume fractions or mass fractions) is represented by a point whose vertical distances to the sides BC, AC and AB are a, b and c, respectively. This method of graphing a function is based on the property of an equilateral triangle that the sum of the vertical distances from a point to the sides of an equilateral triangle is a constant equal to the height of the triangle.

Figure 7.13 shows the phase boundaries for the system chloroform–acetic acid–water at 291 K. Chloroform and water are almost immiscible at this temperature as is indicated by the large two-phase region on the $CHCl_3$–H_2O line. Acetic acid, on the other hand, is completely miscible with both water and chloroform. On adding acetic acid to a two-phase

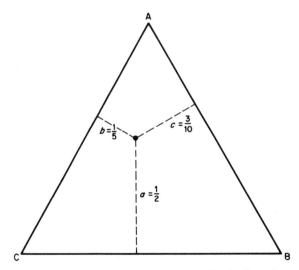

Figure 7.12. Triangular coordinate system. The point indicated has coordinates $(\frac{1}{2}, \frac{1}{5}, \frac{3}{10})$ being the fractional compositions of A, B and C respectively

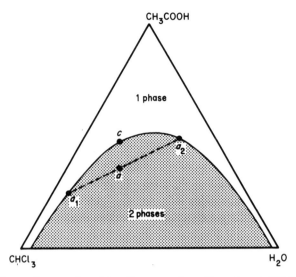

Figure 7.13. Phase diagram for chloroform–acetic acid–water at 291 K. The dotted line indicates a tie line of the two-phase region; a_1 and a_2 indicate the compositions of the two liquid layers in a mixture whose total composition is a

mixture of chloroform and water, the acetic acid will be distributed between the two liquid layers.

For any mixture whose total composition is indicated by a point in the two-phase region the composition of the two layers is indicated by a tie line passing through that point. It must be noted, however, that such a tie line is not parallel to the H_2O–$CHCl_3$ axis nor are all tie lines parallel to one another. The lever rule (7.6) gives the relative amounts of the two liquid phases. If sufficient acetic acid is added the compositions of the two phases become more similar (i.e. points a_1 and a_2 come together) until they finally coincide at the point indicated by c in the figure. This is called a critical point for the system by analogy with the gas–liquid critical point (when gas and liquid become identical).

The phase boundary in Figure 7.13 can be determined by taking a water–chloroform mixture and titrating with acetic acid until the two layers just disappear (on shaking there is no evidence of turbidity). To find the tie line it is necessary to determine the composition of at least one of the liquid phases.

If one wishes to show the temperature-variation of a ternary system it is necessary to take a vertical stack of diagrams such as Figure 7.13 with temperature as the vertical axis. With this extra dimension the boundary between the single-phase and two-phase regions would be a surface, as indicated in Figure 7.14. Under some conditions there may be an upper critical solution temperature (K in the figure) above which the three components are miscible in all proportions.

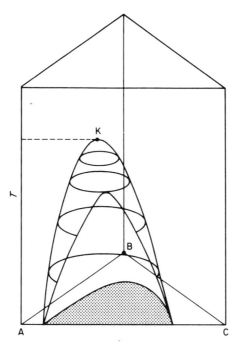

Figure 7.14. Temperature variation of a ternary system showing an upper critical solution temperature K

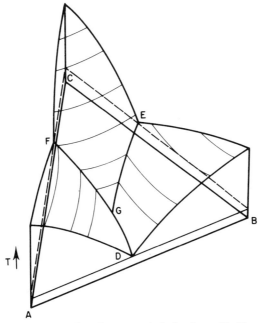

Figure 7.15. A ternary system showing eutectic behaviour. D, E and F are eutectics for the binary systems AB, BC and CA respectively. G is the ternary eutectic

Figure 7.15 shows an interesting ternary system showing eutectic behaviour: KF–LiF–NaF would be one example. Each of the binary systems shows a binary eutectic. The ternary eutectic, indicated G in the figure, is an invariant point (at constant pressure) at which three solid and one liquid phase are in equilibrium. For this system the ternary eutectic temperature is lower than each of the binary eutectics.

7.5 PARTITION COEFFICIENTS AND CHROMATOGRAPHY

Chromatography is an extremely powerful tool for the separation of complex mixtures into their components. The various methods under this label depend for their efficiency upon differences in the distribution of the components between a moving phase and a stationary phase. The different components pass through the system in a time which depends on this distribution.

The first techniques to be explored stemmed from the work of Tswett in 1906 who placed a mixture of plant pigments at the top of a powdered chalk column and washed through with petrol. The components were seen as coloured bands moving down the column; hence, the name which has stuck to the general technique in spite of the fact that colour has no special relevance. This particular method is known as elution chromatography, a name which is applied whenever the moving phase is not absorbed or adsorbed (Chapter 12) by the stationary phase. In contrast, in the technique known as displacement chromatography the moving phase competes with the mixture to be separated for residence in or on the stationary phase.

A major advance was made in 1941 by Martin and Synge who noted that the laws governing the distribution of a solute between two liquids were simpler than those for the distribution between a liquid and a solid. They showed that, with liquid for both the moving and stationary phases, sharper component bands could be obtained than with the Tswett technique. The liquid constituting the stationary phase is adsorbed on a finely powdered solid support.

There are three principal methods using packed columns and a moving liquid phase: liquid–solid chromatography involving adsorption of the solute from solution (see Chapter 12) usually with a single-component liquid; liquid–liquid chromatography, usually called partition chromatography; and ion-exchange chromatography. The latter, which will not be discussed further, makes use of groups such as —SO_3H and —CH_2NR_2 attached to the back-bone of a polymer, itself usually in the form of beads.

Solid supports commonly used include silica, cellulose powder and diatomaceous earths such as kieselguhr and celite. The use of cellulose powder is analogous to the use of sheets in paper chromatography. The cellose binds some of the water from the solvent and, in simple terms, the separation can be regarded as being due to the different affinities which the solute has for the bound water phase and the moving mixed-solvent phase. For example, amino acids can be separated from phenol–water, or from butanol–acetic acid–water. In reality adsorption and even ion exchange will also occur in such systems, and the theoretical demarcation between the three types is ill-defined. Paper chromatography can be effective without mixed solvents; water drawn into the pores by capillarity (Chapter 12), and adsorption from solution can, for example, separate inorganic ions, or as is often demonstrated, the components of inks.

Martin and Synge also suggested that a gas could be used for the moving phase, as was demonstrated by James and Martin in 1952. In gas–liquid chromatography, g.l.c., the stationary liquid is supported on an inert solid and the sample to be studied is carried through a column of stationary phase by a carrier gas. In addition to the use of g.l.c. for qualitative and quantitative analysis of the constituents of a volatile mixture, and the separation of (and purification of) components as an adjunct to preparative techniques, thermodynamic quantities can be measured. To elucidate this aspect we must examine the mathematical background to chromatography in more detail.

Partition Equations

If a solute A is distributed between two phases α, β, than at equilibrium its chemical potential in the two phases must be equal,

$$\mu_A(\alpha) = \mu_A(\beta). \tag{7.9}$$

If the chemical potential is referred to some standard state according to expression 6.75, we have

$$\mu_A^\ominus(\alpha) + RT \ln a_A(\alpha) = \mu_A^\ominus(\beta) + RT \ln a_A(\beta), \tag{7.10}$$

whence (dropping the suffix A for brevity)

$$\ln\left(\frac{a(\alpha)}{a(\beta)}\right) = -\frac{(\mu^\ominus(\alpha) - \mu^\ominus(\beta))}{RT}, \tag{7.11}$$

or

$$\frac{a(\alpha)}{a(\beta)} = \exp(-\Delta\mu^\ominus/RT) \equiv K_a. \tag{7.12}$$

K_a is a constant (an equilibrium constant) whose value depends on the difference in the chemical potentials of the standard states in the two phases.

For the practical application of chromatography we are interested in the ratio of the concentrations (C) of the solute in the two phases rather than the ratio of activities. The ratio

$$\frac{C(\alpha)}{C(\beta)} = K \tag{7.13}$$

is called the partition coefficient. For ideal laws of solubility K will be a constant equal to K_a. In practice K is close to being constant for the distribution of most solvents between two liquid phases or between a liquid and a gas phase. However, for the distribution of a solute between a solid and a liquid or a solid and a gas, K is usually far from being constant. This we refer to as a non-linear distribution (a graph of $C(\alpha)$ against $C(\beta)$ is not a straight line) and it leads to broad bands when chromatography is based on such systems (e.g. in the early elution techniques).

We will first consider the partition of a solute between fixed volumes of two immiscible solvents. This is of practical importance in the common extraction technique used in organic chemistry which is carried out in a separating funnel.

We express the compositions arising in 7.13 as mole fractions. A solution of W kg of solute (molecular weight M) in V_α dm^3 of liquid α (density ρ_α molecular weight M_α) has a mole fraction

$$x(\alpha) = \frac{WM^{-1}}{WM^{-1} + \rho_\alpha V_\alpha M_\alpha^{-1}} \simeq \frac{WM_\alpha}{\rho_\alpha V_\alpha M}, \qquad (7.14)$$

the approximation being valid for a dilute solution. We now add V_β dm^3 of liquid β (density ρ_β, molecular weight M_β) and establish equilibrium. The equilibrium mole fractions of solution in the two phases are, from 7.14

$$K = \frac{x(\alpha)}{x(\beta)} = \left(\frac{W_\alpha M_\alpha}{M\rho_\alpha V_\alpha}\right)\left(\frac{M\rho_\beta V_\beta}{W_\beta M_\beta}\right), \qquad (7.15)$$

whence

$$\frac{W_\alpha}{W_\beta} = \left(\frac{K\rho_\alpha M_\beta}{\rho_\beta M_\alpha}\right)\left(\frac{V_\alpha}{V_\beta}\right) = \frac{K'V_\alpha}{V_\beta}. \qquad (7.16)$$

Writing $W_\beta = W - W_\alpha$, expression 7.16 can be rearranged to

$$\frac{W_\alpha}{W} = \frac{K'V_\alpha}{V_\beta + K'V_\alpha}, \qquad (7.17)$$

which gives the fraction remaining in phase α.

If the layer of liquid β is removed and replaced by fresh liquid, there will be a further extraction and the amount of solute then remaining in phase α, $(W_\alpha)_2$, is obtained from 7.17 by replacing W_α/W by $(W_\alpha)_2/W_\alpha$. If this process is repeated r times, the amount remaining in phase α is given by $(W_\alpha)_r$ where

$$(W_\alpha)_r = W\left(\frac{K'V_\alpha}{V_\beta + K'V_\alpha}\right)^r = W\left(1 + \frac{V_\beta}{K'V_\alpha}\right)^{-r}. \qquad (7.18)$$

It can be seen from 7.17 that for efficient extraction the ratio $V_\beta/K'V_\alpha$ should be as large as possible. A point that is less obvious is that if a fixed total volume, V, of solvent is to be used, then it is better to use n fractions

V/n times ($n > 1$), rather than to make a single extraction with the whole volume.

Although apparatus has been developed to carry out automatically the multiple extractions represented by equation 7.18, it is rather cumbersome. However, one can reasonably infer that a continuous process in which the solvent β is replaced by a moving phase which passes down a packed column containing phase α will simulate a multiple extraction with large r. This is the principle behind elution chromatography.

Gas–Liquid Chromatography

The normal method of carrying out g.l.c. is to take a long narrow tube containing an inert packing material such as silica which is used to support an involatile liquid phase such as squalane ($C_{30}H_{62}$). The mobile phase is usually a constant stream of an inert carrier gas such as nitrogen. The sample is injected by syringe at one end of the column and the time taken to give a signal on a detector at the other end is noted. If the sample contains a mixture of compounds then these will travel down the column at different rates and appear separately at the detector as indicated in Figure 7.16.

The concentration profile of solute that appears at the detector depends on the experimental conditions (e.g. speed of the carrier gas) and a detailed analysis is beyond the scope of this book.† In most cases the profile (Y) fits approximately to a gaussian function

$$Y = h \exp(-x^2/2\sigma^2), \tag{7.19}$$

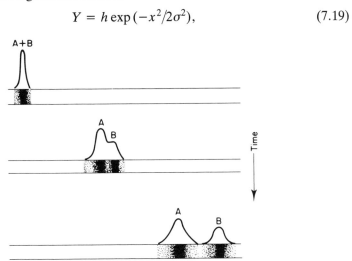

Figure 7.16. Separation of two components in a g.l.c. column. The curves show the concentration profile of the mixture as it passes down the column

†For a more detailed treatment see: J. H. Purnell, *Gas Chromatography*, Wiley, London and New York (1962); J. R. Conder and C. L. Young, *Physicochemical Measurement by Gas Chromatography*, Wiley–Interscience, Chichester (1979).

where h is the peak height, x is the distance along the tube measured from the peak centre and σ is the standard deviation of the distribution. The width at half the peak height is 2.35σ. Measured at the outlet, σ^2 is approximately proportional to the time taken for the solute to pass down the column.

The chromatogram is generally regarded as the solute concentration–time profile shown by the detector (a device which measures the sample concentration in the gas at the outlet). The retention time t_R is the time taken for a sample injected into the column to appear as a peak at the outlet. This time is inversely proportional to the average flow rate of the carrier gas in the column (f_{av}). Solutes can be differentiated either by their retention time or by the volume of carrier gas that is passed in this time which is called the retention volume and is defined by

$$V_R = t_R f_{av}. \tag{7.20}$$

A marker which is insoluble in the stationary phase will pass down the column in a time t_R^0 and have a retention volume V_R^0. The time for a particular solute to appear at the detector will be t_R^0 divided by the fraction of time that the solute molecules spend in the gas phase. If n_g and n_l are the number of moles of solute in the gas and liquid phases then the fraction of time spent in the gas phase is $n_g/(n_g + n_l)$. Although n_g and n_l vary at different places and different times along the tube this fraction will be a constant depending on the partition coeffcient for the solute in the two phases. From this argument we have

$$t_R = t_R^0 \left(\frac{n_g + n_l}{n_g} \right). \tag{7.21}$$

It follows that the ratio of molecules in the gas and liquid phases is given by

$$\frac{n_l}{n_g} = \frac{t_R - t_R^0}{t_R^0} = \frac{V_R - V_R^0}{V_R^0} = \frac{V_R'}{V_R^0}. \tag{7.22}$$

V_R' is called the net retention volume of the sample. The relationship between the three retention volumes we have defined is shown in Figure 7.17.

We define the partition coefficient of the solute by the ratio of concentrations expressed as moles per unit volume

$$K = \frac{C_l}{C_g} = \frac{n_l V_g}{n_g V_l}. \tag{7.23}$$

V_l is the volume of liquid in the column and V_g the volume of gas, which can be equated to V_R^0. Thus combining 7.22 and 7.23 we have

$$K = V_R'/V_l, \tag{7.24}$$

hence K can be determined by measuring the net retention volume and knowing the volume of liquid in the column.

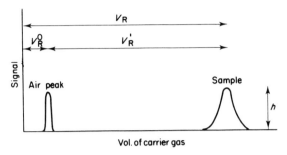

Figure 7.17. Definition of the retention volume and net retention volume in a chromatogram. The carrier gas volume is taken at the temperature and the average pressure in the column

Retention volumes are measured by a gas-flow meter external to the g.l.c. column and appropriate corrections for differences in temperature and pressure between meter and column must be made. An important factor is that the pressure in the column falls between inlet (P_i) and outlet (P_o). The average pressure is approximately the arithmetic mean of P_i and P_o but a more accurate value can be obtained from the expression

$$\frac{P_{av}}{P_o} = \frac{2}{3}\left\{\frac{(P_i/P_o)^3 - 1}{(P_i/P_o)^2 - 1}\right\} \tag{7.25}$$

It has already been said that the partition coefficient is not equal to the equilibrium constant K_a (7.12) unless the gas and liquid phases obey ideal laws. Non-ideality is more pronounced in the liquid phase and can be expressed by the relationship (6.68 and 6.70)

$$p = \gamma p^* x \tag{7.26}$$

where p is the partial pressure of the solute in the gaseous phase, p^* the vapour pressure of the pure solute, x the mole fraction of solute in the liquid and γ the activity coefficient. It should be recalled that in the limit of x approaching zero γp^* is equal to the Henry's law constant (6.72). The limiting value of γ at infinite dilution can, in particular, be measured accurately by g.l.c.

The molar concentration of solute in the gaseous phase is p/RT and that in the liquid phase is $\rho_1 x/M_1$ where ρ_1 and M_1 are the density and molecular weight, respectively, of the pure liquid stationary phase. It follows from 7.23, 7.24 and 7.26 that

$$\frac{V'_R}{V_1} = \frac{RT\rho_1}{\gamma p^* M_1} \tag{7.27}$$

and introducing the total mass of solvent in the column, m, we have

$$V'_R = \frac{RTm}{\gamma p^* M_1}. \tag{7.28}$$

The net retention volume is proportional to the weight of stationary liquid phase in the column and it is therefore customary to tabulate retention volumes per unit weight of solvent, a quantity called the specific retention volume.

For more accurate work, non-ideality in the gaseous phase must be allowed for, and in fact g.l.c. has proved to be an efficient method of measuring second virial coefficients (3.63) of imperfect gases, particularly the contribution to B arising from the interaction of the solute with the carrier gas. The appropriate equations for this analysis can be found in more advanced texts (see footnote on page 155).

It is possible to use g.l.c. for determining a wide range of physical properties. Although the method is not always one of high accuracy it has the great merit of requiring only very small amounts of material and it is not even neccessary that the material be pure.

We have already seen that measured retention volumes will allow a determination of the activity coefficients of the liquid phase. There have been many studies using homologous series (e.g. n-alkanes) and a constant liquid phase, which show that $\log V_R'$ is a linear function of the number of carbon atoms in the molecule. As p^* in 7.28 is a smoothly varying function of the number of carbon atoms it is concluded that γ also varies smoothly in the series. Table 7.1 shows some values for the infinite-dilution activity coefficients of n-alkane mixtures that have been obtained in this way. As the n-values for solute and solvent approach one another the mixture becomes more ideal ($\gamma \to 1$). Several theories of the non-ideality of n-alkane mixtures have been proposed which give quite good agreement with the experimental data.

Inverse Gas Chromatography

Normal (or direct) gas chromatography employs a known stationary phase in order to determine certain properties of an unknown gas phase; in inverse gas chromatography, the situation is reversed and its use enables many

Table 7.1. Infinite-dilution activity coefficients for n-alkanes (C_nH_{2n+2})

n-solute \ n-solvent	22	24	28	30	32	34	36
4	0.805	0.739	0.693	0.681	0.639	0.619	0.594
5	0.816	0.753	0.698	0.680	0.654	0.631	0.606
6	0.859	0.801	0.736	0.712	0.689	0.665	0.639
7	0.893	0.831	0.768	0.743	0.721	0.696	0.672
8	0.927	0.854	0.796	0.772	0.753	0.728	0.701
10	0.950	0.895	0.845	0.830	0.817	0.782	0.751

The data are from J.F. Parcher and K.S. Yun, *J.Chromatog.*, **99**, 193 (1974) are based on measurements for column temperatures between 353 and 393 K.

thermodynamic properties of the solid or the gas–solid system to be evaluated. By monitoring the gas-chromatographic behaviour of a chosen gas or vapour on the solid of interest, it is possible to evaluate activity coefficients, partial molar enthalpies of dilution (and other partial molar quantities), solubility parameters, adsorption isotherms, and specific surface areas; it is also possible to identify phase transitions in the solid. The technique of inverse gas chromatography is particularly useful for polymers (see page 261).

Chapter 8

Polar Liquids

8.1 DIELECTRIC PROPERTIES

If two charges q_1 and q_2 are separated by a distance R in a vacuum their energy of interaction is given by Coulomb's law

$$E = \frac{q_1 q_2}{4\pi\varepsilon_0 R},$$ (8.1)

where ε_0 is the vacuum permittivity which has the value $8.854 \times 10^{-12} \, \text{J}^{-1} \, \text{C}^2 \, \text{m}^{-1}$. If the charges are not in a vacuum, the electrostatic energy is modified by the medium; the molecules of the medium will be oriented and polarized by the two charges and their energy of interaction will be reduced.

If the distance R is large compared with the dimensions of molecules in the medium then the medium can be treated as if it were a uniform material whose electrical properties are characterized by a parameter ε_r called the relative permittivity (earlier called the dielectric constant). In this case the interaction energy of the two charges is given by

$$E = \frac{q_1 q_2}{4\pi\varepsilon_0 \varepsilon_r R}.$$ (8.2)

Water has one of the highest values for ε_r of all liquids and this is a key factor in its ability to solubilize ionic solids.

The relative permittivity is a macroscopic parameter that has no precise meaning for interactions over distances which are of the order of molecular dimensions. For example, if two charges are separated by a few angstroms so that only one or two water molecules could be accommodated in the space between them, then one cannot calculate their interaction energy by expression 8.2, because the intervening water molecules cannot be modelled by a continuous dielectric. Expression 8.2 will at best give an order of magnitude value for the energy. However, although ε_r is a macroscopic parameter its value is, as we shall see, dependent on properties of the molecules which make up the dielectric.

Table 8.1 shows the relative permittivities of some common liquids. The most striking conclusion to be drawn from this table is that dipolar molecules give liquids with much higher relative permittivities than non-polar molecules, and that hydrogen-bonded liquids have exceptionally high values. It is this relationship between ε_r and molecular properties that we now explore.

The classical theory of dielectric behaviour originated in a suggestion of Faraday that in the presence of an electric field the molecules making up the dielectric suffer a displacement of charge and acquire an induced dipole moment. The mathematical consequences of this were examined by Clausius and Mosotti, and their conclusions are embodied in the Clausius–Mosotti law

$$\left(\frac{\varepsilon_r - 1}{\varepsilon_r + 2}\right)V_m = P_m = \frac{\alpha \mathcal{N}}{3\varepsilon_0}, \tag{8.3}$$

where V_m is the molar volume, α is the polarizability of the molecule and P_m is a characteristic property of the dielectric called the molar polarization. For molecules of low relative permittivity P_m is found to be approximately independent of pressure and temperature. It has also roughly the same value in the solid, liquid and gaseous states. This is not true for materials of high relative permittivity.

The relative permittivity is determined by measuring the capacitance of a condenser in an alternating current circuit with (C) and without (C_0) the dielectric between the plates, and taking the ratio

$$\varepsilon_r = C/C_0. \tag{8.4}$$

ε_r is a parameter which depends on the frequency of the electric field, but

Table 8.1. Relative permittivities, ε_r, of some liquids, and dipole moments, μ, of their gaseous molecules

	T/K	ε_r	μ/D
Ar	82	1.54	0
CCl_4	298	2.23	0
C_6H_6	298	2.27	0
$C_6H_5NO_2$	298	34.8	4.22
CH_3OH	298	32.6	1.70
$CHONH_2$	293	109	3.73
HF	200	175	1.82
NH_3	195	25	1.47
HCN	293	115	2.92
H_2O	273	87.9	
	298	78.4	1.85
	373	55.6	

below about 10^6 Hz its value generally changes very little, and it is this low-frequency value that is commonly called the static relative permittivity.

From the electromagnetic theory of light, Maxwell deduced that for a transparent medium the relative permittivity should equal the square of its refractive index

$$\varepsilon_r = n^2 \tag{8.5}$$

and it was shown by Lorenz and Lorentz that the molar refraction, defined by

$$\left(\frac{n^2 - 1}{n^2 + 2}\right) V_m = R_m, \tag{8.6}$$

is independent of temperature. Refractive index also varies with frequency, and for polar molecules it is only at high frequencies that R_m and P_m become equal, that is, the Maxwell relation holds. For non-polar molecules they are approximately equal over the whole frequency range providing that the wave-length at which n is determined is not close to an absorption band.

The reason for the difference between R_m and P_m is that there are two contributions to P_m, one from the alignment of permanent dipoles in the electric field and the other from the creation of induced dipoles in the molecules; only the second of these contributes to the refraction. For non-polar molecules there is no dipole alignment, and for polar molecules at high frequency(above $\sim 10^{12}$ Hz) the electric field is alternating too rapidly for the dipoles to reach an aligned position.

In most solids the molecules are not free to rotate and hence there is no dipole alignment contribution to the polarization: plastic crystals are an exception to this rule.

The Debye Equation

The two contributions to P_m were first recognized by Debye and encompassed in the Debye equation for the polarization which we shall now derive. The energy of an electric dipole μ in a uniform field $\mathscr{E}$ is

$$E = -\mu.\mathscr{E} = -\mu\mathscr{E} \cos \theta, \tag{8.7}$$

where θ is the angle between the dipole and field vectors. The average value of the dipole moment in the direction of $\mathscr{E}$ can be obtained by weighting the dipole component along $\mathscr{E}$ by the Boltzmann factor, $\exp(-E/kT)$, and averaging over all angles

$$\bar{\mu} = \frac{\int_0^\pi \mu \cos \theta\, e^{(\mu\mathscr{E}\cos\theta)/kT} \sin \theta\, d\theta}{\int_0^\pi e^{(\mu\mathscr{E}\cos\theta)/kT} \sin \theta\, d\theta}. \tag{8.8}$$

The integrals in 8.8 can be evaluated by making the substitution

$$x = (\mu\mathscr{E}\cos\theta)/kT, \tag{8.9}$$

which leads to the result

$$\bar{\mu} = \mu\mathscr{L}(x), \tag{8.10}$$

where $\mathscr{L}(x)$ is called the Langevin function

$$\mathscr{L}(x) = \frac{e^x + e^{-x}}{e^x - e^{-x}} - \frac{1}{x}. \tag{8.11}$$

At normal temperatures and field strengths x is very much less than unity and on expanding the exponentials in 8.11 one finds that $\mathscr{L}(x) \approx x/3$. With this approximation the average dipole moment is

$$\bar{\mu} = \mu^2\mathscr{E}/3kT. \tag{8.12}$$

We must add to this average dipole moment the contribution arising from the polarization of the molecule in the electric field. Expression 2.7 relates the induced dipole moment to the field strength. For non-spherical systems $\boldsymbol{\alpha}.\mathscr{E}$ is a function of the molecular orientation in the field and an appropriate Boltzmann average should be taken. This complication is usually ignored as its effect is small, and hence the average value of the dipole moment of a molecule arising from the permanent and induced moments is

$$\bar{\mu} = \left(\alpha + \frac{\mu^2}{3kT}\right)\mathscr{E}. \tag{8.13}$$

It can be seen from 8.13 that the function $(\alpha + \mu^2/3kT)$ behaves as an effective polarizability for field frequences which are sufficiently low that the Boltzmann-averaged dipole alignment can be established. This polarizabiliy when multiplied by Avogadro's number can be equated to the molar polarization P_m. Using the Clausius–Mosotti expression, 8.3, we arrive at the Debye equation

$$P_m = \left(\frac{\varepsilon_r - 1}{\varepsilon_r + 2}\right)V_m = \frac{\mathscr{N}}{3\varepsilon_0}\left(\alpha + \frac{\mu^2}{3kT}\right). \tag{8.14}$$

The Debye equation holds very well for gases, although the relative permittivities need to be measured very accurately, as they are close to unity, for the equation to be confirmed. Figure 8.1 shows a graph of the molar polarization against T^{-1} for HCl, HBr and HI. The molecular dipole moments and polarizabilities given in Table 8.2 have been obtained from the slopes and intercepts of the lines. Dipole moments can be measured more accurately from the Stark splitting of microwave lines but it can be seen from Table 8.2 that the values obtained by dielectric measurements are quite accurate.

If one attempts to use the Debye equation to obtain the dipole moment of molecules in a polar liquid the results are generally very unsatisfactory. We

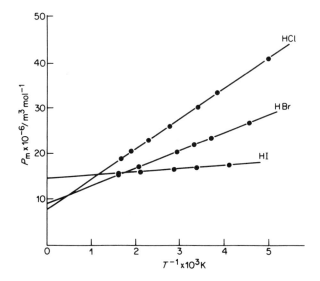

Figure 8.1 Graph of the molar polarization against T^{-1} for the gaseous hydrogen halides. (Data from C. T. Zahn, *Phys. Rev.* **24**, 400 (1924))

Table 8.2. Values of the dipole moment and polarizability determined from the gas-phase relative permittivity (Figure 8.1)

	μ/D	$(4\pi\varepsilon_0)^{-1}\alpha/\text{Å}^3$	$\mu\dagger/D$
HCl	1.05	3.0	1.08 ± 0.01
HBr	0.81	3.5	0.82 ± 0.02
HI	0.39	5.6	0.44 ± 0.02

†Best literature values.

will shortly illustrate this for the case of liquid water. The equation can, however, be used for dilute solutions of polar molecules in non-polar solvents.

For a liquid mixture we can write the total molar polarization as a sum of partial molar polarizations of the components multiplied by their mole fractions

$$P_m = \sum_i \bar{P}_i x_i. \tag{8.15}$$

For mixtures of non-polar liquids an ideal mixing law holds quite well, that is, $\bar{P}_i$ can be taken as the molar polarization $P_{i,m}$ of the pure component i. Although the mixing is not ideal when a polar molecule is one of the components, it is legitimate to assume that $\bar{P}_i$ for the non-polar component

(A) is equal to its pure liquid value when the polar component (B) is in low concentration. Thus the partial molar polarization of the polar component is given by

$$\overline{P_B(x_B)} = (P_m - (1 - x_B)P_{A,m})/x_B, \qquad (8.16)$$

and we are interested in the limit of $\overline{P_B(x_B)}$ as x_B approaches zero.

Figure 8.2 shows $\bar{P}$ for nitrobenzene as a function of mole fraction in solution with hexane. It can be seen that the value extrapolated to infinite dilution is more than three times the molar polarization of pure liquid nitrobenzene.

In principle, the infinite-dilution $\bar{P}$ could be measured as a function of temperature, and dipole moments determined from the Debye equation. In practice this procedure is not often used for two reasons. First, the temperature range over which measurements can be taken for a liquid is often rather small, and there would be a substantial measure of uncertainty in the slope of the Debye graph. Second, the dipole moment of the polar molecule depends to a small extent on its environment, both on the nature of the non-polar solvent and the temperature. Thus P_m may have temperature factors other than those contained in the Debye equation, and these may complicate the situation.

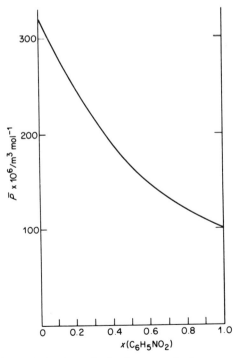

Figure 8.2 The molar polarization of nitrobenzene in hexane solution (293 K) as calculated from expression 8.16 (Data are taken from V. K. Semenchenko and M. Azimov, *Russ. J. Phys. Chem.*, **30**, 1821, 2228 (1956))

166

The normal procedure for deducing a dipole moment from infinite-dilution partial-molar polarization is therefore to obtain an independent measure of the polarizability from refractive index data. Further details are given in more advanced texts.† In the case of nitrobenzene, the dipolar contribution of P_m is so large that we can get an estimate of μ by ignoring the polarizability contribution. From the value of $\bar{P}(0)$ shown in Figure 8.2 we deduce $\mu = 3.9$ D, which is quite close to the measured gas-phase dipole moment of 4.2 D.

Onsager–Kirkwood Equations

If we insert known values of the dipole moment and polarizability of water into the Debye equation (8.14) ($\mu = 1.85$ D, $\alpha/4\pi\varepsilon_0 = 1.48$ Å^3) we arrive at the physically impossible value of 4.2 for the factor $(\varepsilon_r - 1)/(\varepsilon_r + 2)$. Moreover the temperature-dependence of the molar polarization of water is quite unlike the Debye prediction, as can be seen from Figure 8.3. The

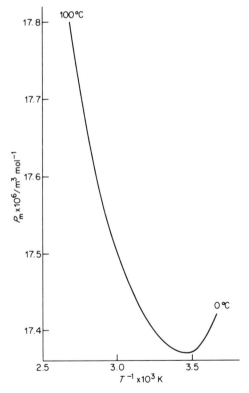

Figure 8.3 The temperature dependence of the molar polarization of liquid water

†J. W. Smith, *Electric Dipole Moments*, Butterworths, London (1955); C. P. Smyth, *Dielectric Behaviour and Structure*, McGraw-Hill, Maidenhead (1955).

water behaviour is not untypical of polar liquids, particularly those whose molecules have high dipole moments.

There are two features of the derivation of the Debye equation that are unsatisfactory. First, no account is taken of the electric field which acts on a molecule due to the polarization of its immediate neighbours. Onsager pointed out that this field, which he called the reaction field, is aligned in the direction of the permanent dipole and is proportional to the permanent dipole moment. Figure 8.4 shows the Onsager field acting on a dipole placed in a spherical cavity. The lines of induction are bent by the field from induced dipoles in the molecules neighbouring the cavity.

The Onsager correction should be applied to dilute solutions of polar molecules in non-polar liquids as well as for pure polar liquids, but in the latter case there is a more important factor to be considered and that is the orientational correlation that exists between the permanent dipole moments of neighbouring molecules. The Boltzmann-averaged alignment factor given by expression 8.12 represents the effect of each dipole acting independently in the field, but if one molecule is aligned in the field, its neighbours will have preferred orientations by virtue of the dipole–dipole electrostatic energy. In the case of water the directional nature of the hydrogen bonds is an additional factor that introduces correlation between the orientations of neighbouring molecules.

There have been several attempts to introduce dipole correlation into the theory of the relative permittivity of polar liquids. The best established theory is that of Kirkwood. He showed that the Onsager field and dipole correlation leads to the expression

$$\left(\frac{(2\varepsilon_r + 1)(\varepsilon_r - 1)}{9\varepsilon_r}\right)V_m = \frac{\mathcal{N}}{3\varepsilon_0}\left(\alpha + \frac{g\mu^2}{3kT}\right). \tag{8.17}$$

A comparison should be made between this and the Debye equation 8.14. The change on the left-hand side is due to the Onsager reaction field. The appearance of the factor g, called the correlation factor, on the right-hand side is due to the dipole correlation. It is defined by

$$g\mu^2 = \boldsymbol{\mu}.\bar{\boldsymbol{\mu}}, \tag{8.18}$$

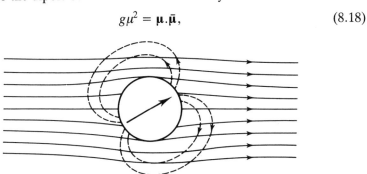

Figure 8.4 The Onsager field arising from a polar molecule in a liquid. The reaction field of the dipole is shown as dashed lines

where $\bar{\mu}$ is the total dipole moment of a molecule and its immediate surrounding molecules.

The Kirkwood theory has been applied to liquid water† with $\bar{\mu}$ calculated from a structure in which each water molecule is hydrogen bonded to four neighbours. Values of g of approximately 2.6 have been obtained with a small decrease as the temperature is raised due to the break-up of the water structure. The value of the relative permittivity obtained from such a calculation is in reasonable agreement with experiment and the dipole moment of a water molecule in the liquid is deduced to be slightly greater than its gas-phase value. One estimate is that $\mu = 2.45$ D which is 32% greater than the gas-phase value.‡

8.2 WATER: THE UNIQUE LIQUID?

Water is the only naturally occurring inorganic liquid on earth and, as it was abundant before life began, it has probably done more than any other chemical substance to shape our biological environment. Water is an essential constituent of animal and plant cells and has a vital influence on the structure of biopolymers. Apart from its role in biochemical reactions, particularly proton transfer, it has a physical role as solvent, lubricant and irrigant.

More chemical and physical sudies have been made on water and its solutions than on any other liquid, but the accumulated data have not proved easy to interpret. Water solutions are very far from ideal in the sense that this term has been used for non-electrolyte mixtures. Even H_2O-D_2O mixtures are slightly non-ideal: the heat of mixing in 1:1 proportions is 1.3 kJ mol^{-1}.

One would expect the properties of water to be most closely paralleled by molecules of similar chemical structure, and the hydrides of the elements neighbouring oxygen in the periodic table obviously bear comparison. NH_3, HF and H_2S are all gases at room temperature due to the fact that the intermolecular energies for these compounds are much weaker than that for water. The binding energies calculated for the dimers of H_2O and HF are quite similar, approximately 24 kJ mol^{-1}, but in the liquid state water molecules can form twice as many hydrogen bonds as can hydrogen fluoride. The hydrogen bond strengths in NH_3 or H_2S are weaker, approximately 16 kJ mol^{-1} and 8 kJ mol^{-1} per bond, respectively, which can be attributed to the fact that nitrogen and sulfur are less electronegative elements than oxygen. Thus the high boiling point of water in comparison with the neighbouring hydrides is due both to the strength of the hydrogen bond and to the number of bonds that can be formed by each molecule.

†J. A. Pople, *Proc. Roy. Soc.*, **A205**, 163 (1951).
‡C. A. Coulson and D. Eisenberg, *Proc. Roy. Soc.*, **A291**, 445 (1966).

Table 8.3. Properties of water and related liquids

	NH₃	H₂O	HF	H₂S	CH₃OH
m.p./K	195	273	184	187	179
b.p./K	240	373	293	212	338
liq.range/K	44	100	109	25	159
ε_r	25(195 K)	79(298 K)	84(273 K)	9(187 K)	33(298 K)

Another molecule that has some features similar to H_2O is CH_3OH. The dimer strengths appear to be very similar for the two species but, like HF, CH_3OH has only one hydrogen atom per molecule that can take part in hydrogen bonding. The relatively high boiling point of methanol in comparison with hydrogen fluoride is due to the fact that CH_3OH gives rise to greater dispersion energies because of its larger size (and hence larger polarizability). A similar fact explains why methylamine has a boiling point 27 K higher than that of ammonia.

As noted in the previous section, the relative permittivity of water is very high due to the high degree of spatial correlation between neighbouring dipoles. However, water is not unique as can be seen from Table 8.3. Liquid HF also has a high relative permittivity, and even higher are those of HCN ($\varepsilon_r = 158$), and formamide ($\varepsilon_r = 109$). The relative permittivity is an important parameter for understanding the solvating power of a polar solvent, but it is not the only factor that must be considered. For example, liquid HCN is generally a poorer solvent for ionic solids than is water, yet its relative permittivity is much higher.

Water becomes supercritical at 374 °C and 220 atm, and under these conditions its density is only one third of its NTP value. In the supercritical fluid the hydrogen bond structure is largely destroyed and water becomes a much better solvent for organic compounds. It is also a poorer ionizing solvent and, in particular, the equilibrium constant for autoionization (K_w), which we refer to in Chapter 10.4, is a factor of 10^5 smaller than that of normal water.

In this brief survey of the properties of water we have shown that it is not exceptional in its physical properties but that does not, of course, detract from the important place it holds in the physics and chemistry of liquids.

8.3 THE STRUCTURE OF ICE

There have been several models of liquid water and aqueous solutions which have invoked some analogy with the structure of ice, and for this reason a brief survey of the solid phases of water is useful.

Whilst the term ice usually refers to the solid phase which is formed when liquid water freezes at atmospheric pressure, an unusually high number of different solid phases of water are known. Figure 8.5 shows the liquid–solid

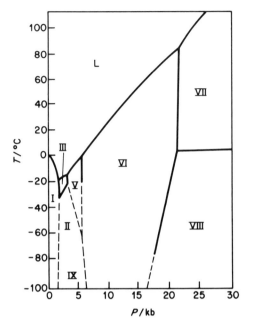

Figure 8.5 The phase diagram of ice (E. Whalley, J. B. R Heath and D. W. Davidson, *J. Chem. Phys.*, **48**, 2362 (1968))

phase diagram deduced by Whalley and co-workers. The solid lines show the measured phase boundaries and the broken lines are estimated boundaries. Some metastable phases are also known including a phase called IV which has been found within the phase V region.

Normal ice, labelled I or Ih, is also called hexagonal ice, a name which has been taken from its lattice structure, which is shown in Figure 8.6. A

Figure 8.6 The lattice structure of ice Ih. Only the positions of the oxygen atoms and the O–H–O hydrogen bonds are shown

cubic modification of ice I is stable below about 153 K. This cannot be prepared by cooling the hexagonal form, but it has been obtained via phases III and II.

Several of the phases have been studied by diffraction techniques and they all appear to have the common feature that each oxygen is hydrogen bonded to four neighbouring oxygen atoms with an O . . . O distance of approximately 2.8 Å. In phase I these neighbours are precisely tetrahedral, but in the high-pressure phases the angles are distorted so that some of the distances between non-bonded oxygen atoms are much reduced from their corresponding phase I distances.

From neutron dffraction studies it has been deduced that the hydrogen atoms lie along or very close to the O . . . O axes. If the water molecules in phase I had the same bond angle as in the gas phase (104.5°) this would imply that the hydrogen atoms were about 2° off the O . . . O axis, but experimental measurements cannot fix their positions to this accuracy.

It can probably be appreciated from Figure 8.6 that normal ice has a rather open lattice structure, and the unusual fact that liquid water at its freezing point is less dense than ice can be attributed more to the low density of ice than to the high density of water. A reasonable interpretation of the fact that at atmospheric pressure water reaches its maximum density at 277 K (4 °C) is that between 273 and 277 K (0 and 4 °C) the structure of liquid water is gradually breaking away from a short-range order similar to that in phase I and moving to a more closely packed structure. Above 277 K the increasing kinetic energy of the molecules will lead to a reduction in density.

Residual Entropy

Bernal and Fowler were the first to postulate that all ice structures conformed to the rule that each oxygen is linked to four other oxygen atoms through hydrogen bonds, with two OH distances being short (roughly the same distance as in a free water molecule) and two being long. However, this still leaves the possibility of having a disordered hydrogen atom sublattice, as illustrated by the two-dimensional model shown in Figure 8.7. Pauling was the first to calculate the residal entropy arising from this disorder.

In a lattice with N water molecules there are $2N$ O . . . O bonds and in each of these the hydrogen atoms have two possible positions. If we consider the four hydrogen atoms surrounding any one oxygen there are sixteen distinct arrangements, but only six of these satisfy the Bernal–Fowler rules: others correspond to the structures H_4O^{2+}, H_3O^+, OH^- and O^{2-} which are energetically less favourable. Thus the total number of hydrogen atom configurations for N molecules is

$$W = 2^{2N}(6/16)^N = (3/2)^N. \tag{8.19}$$

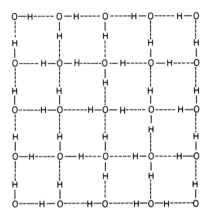

Figure 8.7 A schematic representation of the hydrogen atom disorder in ice

The residual entropy per mole is therefore

$$S = k \ln W = R \ln (3/2) = 3.37 \, \text{J} \, \text{mol}^{-1}. \tag{8.20}$$

This value agrees closely with that deduced from thermal data, and this agreement confirmed the hypothesis (prior to its confirmation by neutron diffraction) that the hydrogen atoms were not placed at the centres of the O . . . O bonds, but were closer to one oxygen than the other.

One would expect a disordered phase to show a phase change to an ordered phase when the temperature is lowered. However, there is no evidence for a phase equivalent to ice Ih with an ordered hydrogen sublattice. In contrast the high pressure forms VII and VIII do appear to be related in having the same oxygen sublattice (body-centred cubic) but with the hydrogen atoms disordered in VII and ordered in VIII. The transition temperature is approximately 280 K.

8.4 THE WATER INTERMOLECULAR POTENTIAL

Because water is such an important and unusual liquid, a great deal of effort has been made to explain its properties from the features of its intermolecular potential. Numerous functions to represent the potential can be found in the literature and their development provides an excellent illustration of advances that have occurred in experimental and computational techniques.

Two distinct approaches to the problem of finding the best potential have been taken. The first was aimed at the exact form of the water dimer potential, and the second was aimed at the best empirical dimer potential that would reproduce the properties of liquid and solid water. These two potentials will only be identical if the total potential function for the liquid and solid phases can accurately be represented by a sum of water dimer functions. If in these bulk phases there are significant contributions to the potential from water molecules interacting three or more at a time, then the empirical dimer potential must contain some average contribution of these higher-order terms.

It is to be expected that the many-body contributions to the potential will be more important for liquid water than for a simple atomic liquid such as argon, or even a molecular liquid made up of non-polar molecules such as liquid nitrogen. The reason for this is that the induction and charge-transfer energies are more important for polar than for non-polar molecules and, as stated in Table 2.3, these energies are not pairwise additive.

McDonald and Klein,† for example, concluded from their computer simulation of liquid water that there was little hope of reconciling the properties of the liquid and gaseous states on the basis of a single water dimer potential. The radial distribution function in the liquid which is computed from a pair potential derived from gas-phase data is in rather poor agreement with the experimental radial distribution function obtained from X-ray data.

It is possible to calculate by quantum mechanics the pair and many-body potentials for small clusters of water molecules and hence to deduce the size of the many-body terms for bulk water. Clementi and co-workers‡ have calculated the energies of clusters of three and four water molecules in many different configurations and have concluded that, whereas the dimer binding energy is approximately $22 \, \text{kJ mol}^{-1}$, the three-body energies typically give an additional stabilization of $7 \, \text{kJ mol}^{-1}$ and the four-body $2 \, \text{kJ mol}^{-1}$. Model potentials have been derived which include many-body terms and, when used in computer simulation of bulk water, give radial distribution functions $g(r)$ which agree quite well with those deduced from X-ray and neutron diffraction studies.

It can be anticipated that future experimental studies of water clusters formed by nozzle-beam expansion will provide further insight into the nature of the water–water potential and the structure of liquid water. Calculations using the best currently available potentials have indicated that clusters of four water molecules (a square array of oxygen atoms linked by hydrogen bonds), and eight water molecules (a cubic array of oxygen atoms linked by hydrogen bonds), are particlarly stable.§ The role which such clusters play in the structure of liquid water is an interesting question.

Evidence for the nature of the exact water dimer potential comes from a variety of sources, particularly measurements on gaseous water. The first such quantity to be employed seriously was the second virial coefficient (3.64). For a monatomic gas the second virial coefficient generally provides

---

† I. R. McDonald and M. L. Klein, *Faraday Disc.*, **66**, 48 (1978). The published discussion of this paper contains several points relevant to this conclusion.
‡ See, for example, U. Niesar, G. Corongin, E. Clementi, G. R. Kneller and D. K. Bhattacharya, *J. Phys. Chem.*, **94**, 7949 (1990).
§ A. Vegiri and S. C. Farantos, *J. Chem. Phys.*, **98**, 4059 (1993).

useful information about the attractive region of the potential but for molecules B values are less useful. Molecular potentials are not just functions of the distance between molecules but they also depend on the relative orientations of the molecules. Second virial coefficients depend only on angular averages of the pair potentials, and hence can give no direct information about the angular features of the potential. Scattering cross-sections measured in low-energy molecular beams can in principle provide information on the angle- and distance-dependence of the potential, but for a system as complicated as the water–water potential such experiments have not as yet given much information.

The most important experimental evidence which exists at present for the intermolecular potentials between small molecules comes from the application of microwave and radio frequency spectroscopy to molecular dimers. These dimers (together with other small clusters) can be formed by expanding a gas through a pinhole nozzle into a vacuum (the so-called supersonic beam). The emerging molecules are characterized by very low rotational and vibrational temperatures (a few degrees kelvin), and as a result the population of clusters held together only by van der Waals bonds is high.

Dyke and co-workers[†] have studied the water dimer in such an experiment. From the rotational energy levels, using isotopic substitution for both the hydrogen and oxygen atoms, it was possible to deduce that the structure of the dimer is a *trans*-linear complex with the geometry shown in Figure 8.8. With further spectroscopic study it should be possible to deduce some of the force constants for this structure and thus enhance our knowledge of the potential energy surface.

There is at present no experimental value for the dissociation energy of the water dimer. Quantum mechanical calculations on the water dimer potential surface give a geometry which agrees closely with the experimental

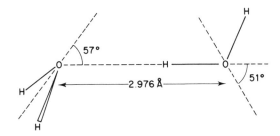

Figure 8.8 The geometry of the water dimer. The symmetry axes of each molecule (shown as dashed lines) are lying in the plane of the paper and make angles with the O–O line as shown. The OHO bond is linear within experimental error. The geometry of each water molecule is assumed to be little changed on dimerization. (J. A. Odutola and T. R. Dyke, *J. Chem. Phys.*, **72**, 5062 (1980))

†T. R. Dyke, K. M. Mack and J. S. Muenter, *J. Chem. Phys.*, **66**, 498 (1977).

geometry, the main difference being that the hydrogen bonding proton deviates slightly from the O–O axis; the O–H bond makes an angle of about 4° with the O–O axis. One therefore has grounds for believing that the dissociation energy calculated for this structure, approximately $22\,kJ\,mol^{-1}$, is not far from the exact value. In the future such calculations are likely to provide the best route for obtaining a complete picture of the potential energy surface of the water dimer and similar species.

The most important feature of the water–water potential in respect to the structure of the solid and liquid phases is probably the angular behaviour as this accounts for the strong preference for tetrahedral coordination. A model potential which most obviously introduces this behaviour was first suggested by Bjerrum; it is based on the location of four point charges within the molecule at approximately tetrahedral directions from the oxygen as centre as shown in Figure 8.9a. Two of the charges are positive and are placed on the hydrogen atoms and the other two are negative and are placed at points which are identified as the centres of density of the lone-pair electrons.

The electrostatic interaction between these point charges on different molecules will lead to tetrahedral coordination with the distance between adjacent molecules being controlled by a repulsive potential which acts either between the centres of the molecules or between individual atoms in the molecules. For example, a potential devised by Ben-Naim and Still-inger† combines a Lennard-Jones potential (2.15) acting between the oxygen atoms with the electrostatic interaction between tetrahedral site charges.

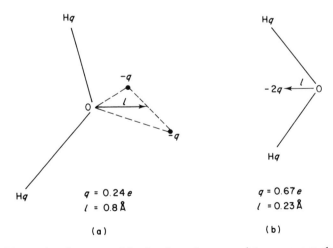

(a)   (b)

Figure 8.9 Two point-charge models that have been used to represent the electrostatic part of the water–water potential. (a) F. H. Stillinger and A. Rahman, *J. Chem. Phys.*, **60**, 1545 (1974). (b) H. Popkie *et al.*, *J. Chem. Phys.*, **59**, 1325 (1973).

†A. Ben-Naim and F. H. Stillinger, *Structure and Transport Properties in Water and Aqueous Solutions*, Ed. R. A. Horne, Wiley-Interscience, London (1972).

There have been several calculations of liquid water properties using this potential or variants of it.

The angular feature of the water–water potential can be introduced through other models than the Bjerrum tetrahedral charges. For example, a potential fitted to quantum mechanical calculations of the water dimer surface has for its electrostatic part been represented by only three point charges, all of which are on the same side of the oxygen atom. This electrostatic model shown in Figure 8.9(b) was combined with exponential repulsion terms acting between all pairs of atoms in the system.† However, most computer simulations of liquid and solid water in the future will be made using potentials that include many-body terms, so that these early potentials illustrated in Figure 8.9 are mainly of historical interest.

8.5 STRUCTURAL MODELS OF LIQUID WATER

Because of its importance as a solvent and its interesting properties, there have been many attempts to establish structural models of liquid water. The properties that have engaged most attention are its anomalous density, its X-ray and neutron-diffraction patterns, its relative permittivity, and its infra-red and Raman spectra.

All models attribute a high degree of ice-like order to cold water at low pressures and attempt to explain its disappearance as either temperature or pressure is raised. The models fall roughly into three classes. The first are mixture models, the simplest of which assumes water to consist of ice-like regions and fluid regions. These regions are not permanent, but move around the liquid through breaking and forming of hydrogen bonds. The relative proportions of the two regions will be a function of temperature and pressure.

In the second class of model, water is considered to be uniform in structure but the hydrogen bonds are distorted in varying degrees from the ideal geometry which exists in ice I. The average degree of bond distortion is a function of temperature and pressure.

The third class of model is based on the view that the bulk of water is ice-like, albeit with distorted hydrogen bonds, but individual water molecules occupy some of the empty space in the ice lattice as interstitial molecules. It is this filling of spaces that is used to explain the anomalous density of water.

Vibrational Spectra

The infra-red and Raman spectra of H_2O, D_2O and their mixtures tend to support either the mixture or the interstitial models. The vibrational

---

†H. Popkie, H. Kistenmacher and E. Clementi, *J. Chem. Phys.*, **59**, 1325 (1973).

frequencies associated with hydrogen motion in liquid water occur in the region of $3500\,cm^{-1}$ for O–H stretch ($2500\,cm^{-1}$ for D_2O), $1600\,cm^{-1}$ for HOH bend and between $1000\,cm^{-1}$ and $300\,cm^{-1}$ for libration. The latter can be thought of as a mixture of hindered rotation of the water molecules and hydrogen bond bending.

If the Raman spectrum of a 1% mixture of H_2O in D_2O is examined (almost all the hydrogen is then in HDO species) the OH stretching band appears to have two components, as shown in Figure 8.10. A dilute mixture of D_2O in H_2O shows, less prominently, a similar two-component profile for the OD stretching band. These experiments suggest that there are two types of environment for the dilute isotope.

The relative intensities of the two components in the Raman bands change as expected with a rise in temperature and only the high-frequency component is evident in high-density supercritical water. The corresponding isotopically-mixed ices also show only a single band which in this case is the low-frequency component shown in Figure 8.11.

Both the mixture and interstitial models involve two types of environment for a water molecule and hence are consistent with the Raman results. The uniform model has a continuous range of environments for the hydrogen atoms and appears to be inconsistent with the results. However, a uniform model may still possess a bimodal distribution of environments (i.e. two peaks of probabiity amongst a broad distribution) and hence cannot be ruled out.

Diffraction

The most important experimental data for judging the validity of any model of liquid water are those from X-ray and neutron diffraction experiments. It is therefore not surprising that a great deal of effort has been made to collect

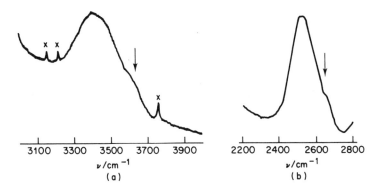

Figure 8.10 Argon-ion-laser Raman spectrum of (a) 1% H_2O in D_2O; (b) 1% D_2O in H_2O at 298 K. The sharp lines marked with a cross are argon lines. The arrows show the weak second components (G. E. Walraten, *Water, a Comprehensive Treatise*, Vol. 1, Chapter 5, ed. F. Franks, Plenum Press, London and New York (1972))

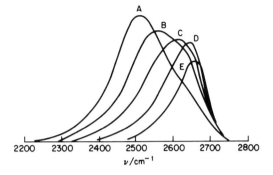

Figure 8.11 High-temperature, high-pressure Raman spectra from 11 mole % D_2O in H_2O. A, 293 K, 100 atm; B, 373 K, 1000 atm; C, 473 K, 2800 atm; D, 573 K, 4700 atm; E, 673 K, 3900 atm (Data of H. Lindner reported in G. E. Walraten, *Water, a Comprehensive Treatise*, Vol. 1, Chapter 5, ed. F. Franks, Plenum Press, London and New York (1972)

such data for a range of temperatures and pressures, and to analyse them to obtain radial distribution functions. An important conclusion from this work is that there is no evidence for large ice-like regions in liquid water. These would be expected to possess long-range order and would give rise to oscillations in the small angle X-ray scattering intensities. No variation has been found in the parameter s (defined by 1.9) in the region $0.04 \text{ Å}^{-1} < s < 0.5 \text{ Å}^{-1}$ for MoK_α radiation ($\lambda = 0.7107 \text{ Å}$) and this rules out any order existing over more than about 10 Å. A mixture model, if taken to imply large ice-like regions, is therefore untenable.

Figure 8.12 shows the radial distribution function $g(r)$ (defined in Section 1.3) which has been obtained from the Oak Ridge X-ray data. This radial distribution function includes positional correlation for both the oxygen and hydrogen atoms. The X-ray scattering arises predominantly from oxygen atoms but the contributions from O$\cdots$H (12%) and H$\cdots$H (2%) are not negligible.

The prominent peak which is seen in $g(r)$ at all temperatures at approximately 1 Å must be due to the intramolecular O–H distance. The second prominent peak at about 2.9 Å will be due to the nearest O–O distance. It can be seen that this increases from 2.84 Å at 4 °C to 2.94 Å at 200 °C.

The peak in $g(r)$ at 2.9 Å is flanked on both sides by ripples which are thought to be an artefact of the analysis due to termination of the Fourier integral (1.13) at a finite value of s. However, the broad maxima at approximately 4.5 Å and 7.0 Å are physically real and represent average second and third neighbour distances. These peaks gradually disappear as the temperature is raised.

In ice I the shortest O–O distance is 2.76 Å and the second and third O–O neighbours are separated by 4.75 Å and 5.2 Å, respectively. The

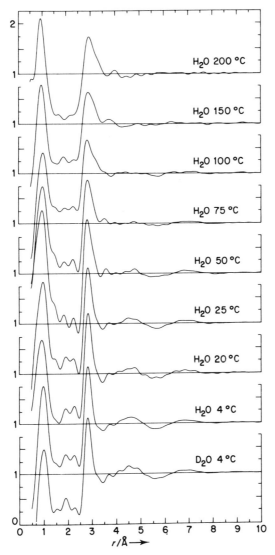

Figure 8.12 The radial distribution function $g(r)$ for water determined from the Oak Ridge data. The figure is taken from A. H. Narten and H. A. Levy, *Water, a Comprehensive Treatise*, Vol. 1, Ed. F. Franks, Plenum Press, London and New York (1972)

second neighbour distances in ice I and liquid water are therefore very similar and we conclude that the average OÔO angle made by an oxygen atom and its first and second neighbours is close to the tetrahedral value (109°) which occurs in ice I. At third neighbours, however, the difference between ice I and liquid water is considerable.

To obtain from experiment the separate radial distribution functions for the different types of atom in water, g_{OO}, g_{OH} and g_{HH}, it is necessary to have data from other scattering experiments in which the ratio of the scattering factors (the parameter A in expression 1.10) for hydrogen and oxygen is different from that pertaining to X-rays. There has been some analysis of electron scattering from thin films of water but the results are controversial, and in any case it is not clear that the internal structure of a thin film is the same as that of bulk water.

Neutron scattering is the obvious technique to use. It has the advantage over X-ray scattering that different isotopes have different scattering factors. Unfortunately, normal hydrogen gives a large amount of incoherent scattering due to its large nuclear spin, and as yet neutron scattering on H_2O has given no structural information. In addition, all the oxygen isotopes have about the same scattering factor and hence using a rare oxygen isotope leads to no new informaion. Thus only D_2O provides useful neutron diffraction data.

To provide a third condition (additional to X-ray and neutron data) on the radial distribution functions and thus allow all three to be determined, Narten assumed that the dominant structure in liquid water was a molecule surrounded tetrahedrally by four other molecules as shown in Figure 8.13. With such a model, if g_{OO} and g_{OH} are known g_{HH} can be determined. The resulting distribution functions for water at 298 K are shown in Figure 8.14. (The intramolecular O–H correlation has been subtracted from g_{OH}.)

Computer Simulation

Figure 8.13 also shows the radial distribution functions calculated by Lie and Clementi by the Monte Carlo method, using a water–water pair potential. The agreement between the diffraction experiment and the computer experiment is very good when one considers that neither is made without assumptions: the diffraction results include the Narten assumption about the

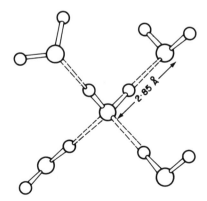

Figure 8.13 The dominant structure in water, which was assumed by Narten in order to separate the total radial distribution function into g_{OO}, g_{OH} and g_{HH}

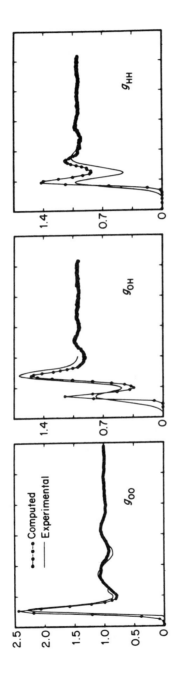

Figure 8.14 The radial distribution functions for water at 298 K determined from X-ray and neutron diffraction data (A. H. Narten, *J. Chem. Phys.*, **56**, 5681 (1972)). The results are compared with those from a Monte Carlo computer simulation (G. C. Lie and E. Clementi, *J. Chem. Phys.*, **62**, 2195 (1976))

dominant water structure and the computer results assume that a pair-additive potential is adequate for liquid water. More recent, and more accurate, neutron scattering data by Thiessen and Narten[†] and by Soper and Phillips[‡] have been analysed by Clementi and co-workers, using the many-body potential referred to earlier (p. 173); there is, in general, excellent agreement between the experimentally derived and theoretically derived radial distribution functions.

A careful analysis[§] of the slow neutron scattering for liquid D_2O has led to the conclusion that the intramolecular $O-D$ distance is 0.983 ± 0.008 Å. This is larger than the value appropriate to an isolated molecule (0.958 Å), but smaller than the distance in ice I, which is generally taken as 1.01 Å although shorter distances have been proposed.

The results of computer simulation suggest that it is naive to adopt any structural model of water in which there is a sharp distinction between molecules which are hydrogen bonded and those which are not. There appears to be a wide range of $O-H\cdots O$ interactions with no clear demarcation between hydrogen bonding and non-hydrogen bonding situations.

Rahman and Stillinger analysed their molecular dynamics results by counting the number of hydrogen bonds that fell within different energy limits. Their results are shown in Table 8.4: E_{HB} is the minimum energy required for a bond to be classified as a hydrogen bond; $\bar{b}$ is the average number of hydrogen bonds terminating at an oxygen atom; N_0 is the fraction of molecules without hydrogen bonds; and N_1 is the fraction of molecules having only one hydrogen bond. It can be seen that if a very low limit is chosen for the hydrogen bond energy then $\bar{b}$ is almost four and there are no free molecules. However, if the highest limit is chosen for the hydrogen bond energy then there is only, on average, one hydrogen bond per molecule and less than half of the molecules have the two hydrogen bonds per molecule found in ice-like structures.

More recent studies[¶] show that the average number of hydrogen atoms around an oxygen is close to four at all temperatures; the average number of

Table 8.4. Analysis of hydrogen bond structure in liquid water from molecular dynamics computer simulation

E_{HB}/kJ mol^{-1}	> 8.87	> 12.80	> 16.47	> 20.27
$\bar{b}$	3.88	3.14	2.26	1.18
N_0	0	0.003	0.041	0.249
N_1	0.003	0.029	0.180	0.415

A. Rahman and F. H. Stillinger, *J. Amer. Chem. Soc.*, **95**, 7943 (1973).

[†]W. E. Thiessen and A. H. Narten, *J. Chem. Phys.*, **77**, 2656 (1982).
[‡]A. K. Soper and M. G. Phillips, *Chem. Phys.*, **107**, 47 (1986).
[§]J. G. Powles, *Molec. Phys.*, **42**, 757 (1981).
[¶]G. Corongin and E. Clementi, *J. Chem. Phys.*, **97**, 2030, 8818 (1992).

oxygen atoms around another oxygen is four at low temperatures but five at high temperatures, the change coming at approximately 320 K.

Although these results give no simple answer to the question of how many hydrogen bonds there are in liquid water they give one unambiguous result. Whatever lower limit is placed on the hydrogen bond energy the distribution function for the number of hydrogen bonds per molecule is always singly peaked as shown in Figure 8.15. Any mixture model for liquid water would probably give a bimodal distribution for hydrogen bonds arising from two different types of environment.

In reviewing the evidence for the structural models of water in 1972. H. S. Frank came down marginally in favour of the interstitial model.[†] The uniform models he thought to be contrary to the infra-red and Raman data that we have already referred to. Since that date the evidence of computer simulation has come down strongly in favour of the uniform model with no evidence for two distinct types of molecule in the liquid. The question of whether this conclusion conflicts with the vibrational behaviour remains to be answered. A recent computer simulation of the dynamical motion in liquid water should provide valuable new evidence on this matter.[‡]

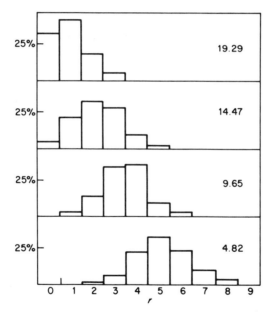

Figure 8.15 Distribution of molecules according to the number of hydrogen bonds N_r in which they engage. The inset numbers are the minimum energies (kJ mol^{-1}) required to count as hydrogen bond. These numbers were obtained using a slightly different potential to that which produced the results of Table 8.3. (A. Rahman and F. H. Stillinger, *J. Chem. Phys.*, **55**, 3336 (1971))

[†]H. S. Frank, *Water, a Comprehensive Treatise*, Vol. 1, Chapter 14, Ed. F. Franks, Plenum Press, London and New York (1972).
[‡]G. Corongin and E. Clementi, *J. Chem. Phys.*, **98**, 4984 (1993).

There are, of course, other properties of water which provide a test of structural models or computer simulation. Relative permittivities are referred to in Section 8.1. Transport properties, heat capacity, nuclear magnetic resonance and many others have received attention. Readers are directed elsewhere for more information, and in particular to the volumes titled *Water, a Comprehensive Treatise*, edited by F. Franks, which have already been referred to.

8.6 NON-AQUEOUS POLAR SOLVENTS

Polar solvents are important because they solubilize polar molecules to produce electrolyte solutions. The physical and chemical properties of ions can therefore be studied in a homogeneous medium. Water is, of course, by far the most important polar solvent and, in respect to the range of compounds which will dissolve in it, is one of the strongest solvating liquids. However, there are many salts which are virtually insoluble in water, AgCl for example, yet other solvents can be found to dissolve them.

A comprehensive theory of ion solvation in a polar medium must recognize that there are at least two contributions to the solvation energy. The first is the electrostatic energy which may be interpreted macroscopically through the relative permittivity or microscopically through the interaction of the ion charge and the dipole moments of the solvent molecules. The second contribution is from covalent bonding between ion and solvent molecules, which is most simply described in terms of charge-transfer forces. These were discussed briefly in Chapter 2.

The importance of the two effects is most clearly seen by comparing the solubilities of alkali-metal and transition-metal salts in water and liquid ammonia. Water has a larger dipole moment and relative permittivity than ammonia and will therefore give a larger electrostatic binding. On the other hand, ammonia forms stronger covalent bonds with cations because its ionization potential is lower (easier electron transfer from solvent to cation)

Table 8.5. Properties of common polar organic solvents

	m.p./K	b.p./K	ε_r(K)	μ/D
1,4-Dioxane	285.0	374.5	2.21 (298)	0.45
Diethylether	156.9	307.8	4.43 (293)	1.15
Pyridine	231.6	388.5	12.4 (294)	2.37
Acetone	178.5	329.5	20.7 (298)	2.69
Ethanol	159.1	351.5	24.6 (298)	1.66
Nitromethane	244.6	374.4	35.9 (303)	3.56
Sulfolane	301.7	560.5	43.3 (303)	4.81
Dimethylsulfoxide	291.7	462.2	46.7 (298)	3.9

Data from J. A. Riddick and W. B. Bunger, *Techniques of Chemistry*, 3rd Edn, Vol. II, Organic Solvents, Wiley–Interscience, London (1970).

Table 8.6. Properties of common inorganic polar solvents

	m.p./K	b.p./K	ε_r	μ/D	$-\log K_i$
HF	189.8	292.7	80 (273)	1.83	9.7
H_2O	273.16	373.16	78.3 (298)	1.85	14.0
NH_3	195.5	239.8	22.7 (223)	1.47	~30
H_2SO_4	283.6	543 (with decomp.)	~110	—	3.6
SO_2	197.7	263.1	20 (233)	1.62	—

Most data from *Chemistry of Nonaqueous Solvents*, Ed. J. J. Lagowski, Academic Press, London and New York (1966–).

and because its lone-pair electrons are spatially more suitable for overlapping cation acceptor orbitals.

For alkali-metal cations and for anions the electrostatic energies are more important than covalent energies, but for transition-metal ions and other heavy-metal cations like Ag^+ the covalent energies are probably dominant. Thus Na_2SO_4 is soluble in water but insoluble in liquid ammonia, whereas AgCl is insoluble in water but soluble in liquid ammonia.†

There are several reasons, apart from solubility, for using a polar solvent other than water in solution studies First, a different temperature range for the liquid state may be required. For example, HF, SO_2 and NH_3 all have melting points around 190 K and are valuable liquids for handling compounds which are unstable at room temperature. In contrast, molten salts can have temperatures above 1000 K and will dissolve metals and normally insoluble solids such as SiO_2.

Further interest in non-aqueous polar solvents arises from the variation that can be obtained in the chemical stability of solutes. For example, alkali metals react with water but in liquid ammonia they form stable solutions with strong reducing properties. In Chapter 10 we shall examine chemical equilibria in solution and, in particular, see how acid–base and redox equilibria depend on the nature of the solvent.

Molten Salts

Molten salts are extremely important in metallurgical processes because most methods of extracting metals from their ores by chemical or electrolytic reduction are carried out in the molten state. In addition, there has been a growing interest in preparative inorganic chemistry in molten salt media.

The melting points of pure salts range from about 500 K to well over 1500 K. $LiNO_3$, for example, melts at 527 K, and K_2SO_4 at 1342 K. Many organic salts have much lower melting points (e.g. Bu_4NI, m.p. = 417 K) and with eutectic mixtures values below 400 K can be obtained (e.g. an $AlCl_3$–NaCl–KCl eutectic melts at 362 K).

---

†A more detailed discussion of the solvation of ions in water will be given in Section 9.1.

The liquid range is typically 500–1000 K although some salts with oxyanions or organic cations will decompose before the boiling point is reached.

Most molten salts have a high electrical conductivity and obey Faraday's laws of electrolysis. It can be deduced from this that in the molten state they are largely composed of ions. There are exceptions; $HgCl_2$, for example, has a low conductivity and a very small liquid range (26 K) indicating that the melt is composed mainly of uncharged species.

From X-ray or neutron diffraction experiments it is possible to deduce a radial distribution function for the ions in the melt. If one makes use of isotopic enrichment of samples then it is possible to obtain separately the distribution functions for distinct pairs in the mixture. Figure 8.16 shows the results of such an analysis on molten NaCl. The coordination numbers deduced by integrating the first peaks are 5.8 ± 0.1 for unlike ions and 13.0 ± 0.5 for like ions. These numbers are quite close to those pertaining to the solid phase, 6 and 12, respectively. Early X-ray measurements had suggested that the unlike ion coordination number was as low as 4, which would explain the large increase in molar volume on melting, but this is now thought to be incorrect.

The computer simulation of ionic melts using the Monte Carlo or molecular dynamics techniques described in Chapter 3 is difficult because of the long-range nature of the Coulomb force. The Coulomb potential of an ion decreases as $1/r$, r being the distance from the centre and for a uniform density the number of interacting ions at distance r is proportional to r^2. Thus the Coulomb energy of an ion with all others of the same charge at distance r is a divergent function.

There is, of course, a divergent interaction also between ions of opposite sign and there is a cancellation between the two at large distances, but how

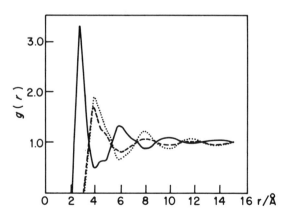

Figure 8.16 The radial distribution functions for NaCl at 1123 K, as determined by neutron diffraction. (F. G. Edwards, J. E. Enderby, R. A. Howe and D. I. Page, *J. Phys. C*, **8**, 3483 (1975)) Solid, Na–Cl; dashed, Na–Na; dotted, Cl–Cl

best to introduce this cancellation has been a controversial matter. Moreover, whatever technique is adopted, one cannot get around the fact that the length of the cell considered in the computer simulation (see Figure 3.4) should be greater than the distance over which pair correlation persists and, as Figure 8.16 shows, this is rather large. In other words, one cannot make a satisfactory computer simulation of an ionic melt without considering a large number of ions in the cell, preferably above 10^3.

The melting of an ionic solid is generally accompanied by a large volume increase, typically about 20%. This has been attributed to the presence of holes in the melt, which in turn would explain the generally high ionic mobility. However, computer simulation does not support the concept of holes in the sense of missing ions in a nearly regular lattice into which neighbouring ions can jump. Instead the concept of 'voids' with a short lifetime ($\sim 10^{-12}$ s) which collapse by the concerted motion of all neighbouring ions, is preferable.

Complex Ions

The phase diagrams of salt mixtures show a variety of patterns. Solid solutions can be obtained if the salts have cations of similar sizes and anions of similar sizes (e.g. LiBr and NaBr). In other cases there is the formation of a simple eutectic (LiF–KF) or compound formation (LiF + CsF → LiCsF$_2$). Complex ions are frequently present in melts of mixed salts, particularly when transition metals are involved, and these can be identified by spectroscopy. For example, the Raman spectrum of KCl–ZnCl$_2$ shows the presence of $ZnCl_3^-$ and $ZnCl_4^{2-}$; the electronic spectrum of VCl$_3$ in a LiCl–KCl eutectic shows the presence of VCl_4^- and VCl_6^{3-}.

Complex ion formation can also be deduced from cryoscopic measurements on mixed salts. Because the freezing point depression is a colligative property, it depends on the number of distinct molecules or ions of solute. Thus expression (6.57) for a solute that dissociates is

$$\Delta T_{\mathrm{m}} = \left(\frac{R T_{\mathrm{m}}^2}{\Delta H_{\mathrm{melt}}} \right) i x \tag{8.21}$$

where x is the mole fraction of solute and i is the number of independent particles per formula unit of solute which differ from those present in the solvent. Thus for KCl in AgNO$_3$, $i = 2$ but for KNO$_3$ in AgNO$_3$, $i = 1$. HgCl$_2$ in AgNO$_3$ gives a value $i = 2$ at infinite dilution and not $i = 3$ which would be obtained for complete dissociation. One possible explanation is that the ions present are $HgCl^+$ and Cl^-. Another possibility is that there is a high proportion of undissociated molecules present in the melt, a point already noted for pure HgCl$_2$. Experience suggests that a great deal of care is needed in interpreting the cryoscopic behaviour of ionic melts as there are large deviations from ideal behaviour even in very dilute conditions.

Mixtures of ionic salts and metals show interesting behaviour. Such mixtures fall roughly into two classes. The first, typified by K in KBr, have a high solubility of the metal in the salt and the conductivity of the mixture is almost entirely electronic as opposed to ionic. It can be concluded that the metal is ionized in the salt medium with the electrons being more or less freely delocalized on the halide anions. The second class, e.g. $Ni + NiCl_2$, has a rather low metal solubility and little increase in conductivity from that of the pure salt. In these systems the metal will reduce the salt to a subhalide, e.g. forming the Ni(I) oxidation state. Ionic melts have in consequence proved valuable for the synthesis of some normally unstable low oxidation states.

Chapter 9

Aqueous Solutions of Electrolytes

9.1 THE HYDRATION OF IONS

One of the most puzzling questions to emerge from a student's first contact with chemistry must be why some inorganic salts are soluble in water and others insoluble. For example, silver fluoride is very soluble, silver iodide is almost insoluble; yet calcium fluoride has a low solubility while calcium iodide is very soluble.

As a general rule, like dissolves like. Polar or ionic solids are more soluble in a polar solvent such as water than in a non-polar solvent such as benzene. A non-polar solid such as naphthalene is more soluble in benzene than in water. In general we can say that if A is a solvent and B a solute then B will dissolve in A only if the A–B interactions are energetically more favourable than the average of the A–A and B–B interactions. Naphthalene is insoluble in water but soluble in benzene, not because water–naphthalene interactions are less stabilizing than benzene-naphthalene, but because water–water interactions are very much more stabilizing than benzene–benzene. A water molecule would much prefer to be adjacent to another water molecule than to a naphthalene molecule. Likewise, most salts do not dissolve in benzene because the salt–benzene interaction is much less stabilizing than the salt–salt interaction.

The solubility of an ionic solid in water therefore results from a balance between the free energy of hydration of its ions on the one hand and the lattice free energy and water free energy on the other. The enthalpic contributions to these free energies are usually the most important terms.

The lattice enthalpy (cohesive energy) of an ionic solid is the energy liberated when one mole of solid is formed from its component ions in the gaseous state. This energy is primarily due to the electrostatic interaction amongst the ions, which in turn depends on the absolute and relative sizes of the anions and cations. Figure 9.1 shows the enthalpy relationship between

190

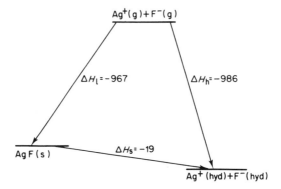

Figure 9.1. Relationship between enthalpies of the lattice, of hydration and of solution for AgF $(kJ \, mol^{-1})$

an ionic solid, its free gaseous ions and the ions in solution. ΔH_1 is the lattice enthalpy, ΔH_h the hydration enthalpy and ΔH_s the enthalpy of solution (all negative in Figure 9.1). A similar relationship exists for other thermodynamic quantities such as entropy and free energy.

Table 9.1 shows the lattice enthalpies and solubilities in water of AgF, AgI, CaF_2 and CaI_2. Because the lattice enthalpy of AgF is greater than that of AgI, yet AgF is the more soluble, we can conclude that the enthalpy of hydration of F^- is greater than that of I^-. Because the lattice enthalpy of CaI_2 is greater than that of AgI, yet CaI_2 is the more soluble, we deduce that the enthalpy of hydration of Ca^{2+} is greater (much greater) than that of Ag^+. However, the hydration enthalpies of Ca^{2+} and F^- are not sufficiently negative to make CaF_2 soluble in water.

For a solid to be appreciably soluble in water, ΔG_s must be negative. A more detailed discussion of the equilibrium between a solid and its ions in solution will be given in Section 10.3. For most soluble salts ΔH_s is also negative, although there are many examples (e.g. NaCl) where it has a small positive value. In this case there is a lowering of the temperature when the salt is dissolved in water.

Most salts dissolve in water with complete dissociation to ions. This is shown by the fact that the conductivity of their solutions per mole of dissolved salt is almost independent of the salt concentration (see Figure 9.3).

Table 9.1. Lattice enthalpies and solubilities (in cold water) of some ionic solids

	$-\Delta H_1/kJ \, mol^{-1}$	Sol./$g \, kg(water)^{-1}$
AgF	967	1820
AgI	889	1.6×10^{-6}
CaF_2	2630	0.016
CaI_2	2074	2090

These salts are referred to as strong electrolytes.† In contrast, a compound like acetic acid is a weak electrolyte. It dissolves in water to give a mixture of ions and undissociated molecules, and the molar conductivity increases rapidly with increasing dilution.

For strong electrolytes it is natural to assume that the thermodynamic parameters for their dilute aqueous solutions can be expressed as a sum of contributions from each of the ions. The formal basis of this assumption will be examined in Section 9.3, particularly the effect that ions have on one another. At this point we just discuss the problem of how the separate ion contributions can be determined.

The enthalpy of hydration of NaCl is $-780 \, \text{kJ mol}^{-1}$ but, as there is always an equal number of Na^+ and Cl^- ions in solution, we have no unique way of dividing the total enthalpy into the separate enthalpies of hydration of the two ions. If the hydration enthalpy is known for any one ion then all others can be found by reference to it. The ion which has received most attention as a standard is H^+. As early as 1919 Fajans proposed the value of $-1100 \, \text{kJ mol}^{-1}$ for its enthalpy of hydration at 298 K and this has stood well against later investigations which have given values between approximately -1000 and $-1250 \, \text{kJ mol}^{-1}$: $-1125 \, \text{kJ mol}^{-1}$ is now commonly accepted as the best value.‡

The Born Equation

One of the earliest attempts to establish a model for ion hydration energies was made by Born. Taking the ion (charge Ze) as a uniformly charged hard sphere of radius r embedded in a fluid of relative permittivity ε_r he obtained the free energy of hydration, per mole, as

$$\Delta G_h = \mathcal{N} \frac{Z^2 e^2}{8\pi \varepsilon_0 r} \left[1 - \frac{1}{\varepsilon_r} \right]. \tag{9.1}$$

Ions are not, of course, hard spheres with definite radii, but effective ionic radii can be deduced from the lattice spacings in ionic crystals. The radius of Na^+, for example, is approximately 0.95 Å; inserting the appropriate parameters into the Born equation gives the result $\Delta G_h(Na^+) = -720 \, \text{kJ mol}^{-1}$. However, the generally accepted value is $-410 \, \text{kJ mol}^{-1}$ so that the Born model is, in absolute terms, not very good.

The deficiencies in the Born equation lie not only in the assumption of a hard-sphere ion, but also in treating the solvent as a continuous dielectric. The hydration free energies of positive and negative ions of the same radius would be the same in the Born model yet we know that the arrangement of

---

†Further evidence, first supplied by Bjerrum, is that the visible and u.v. absorption spectra of strong electrolyte solutions are the sums of spectra associated with cations and anions. Only at high concentrations do bands appear which are attributed to ion pairing.
‡L. G. Hepler and E. M. Wooley, *Water, a Comprehensive Treatise*, Vol. 3, p. 163, Ed. F. Franks, Plenum Press, London and New York (1973).

water molecules around the ions is quite different in the two cases. The equation also leads to the prediction that the solubility of an ionic compound will increase with increasing solvent dielectric constant and this is not always true. However, most developments of the Born model, and there have been several, have just added a correction factor to the ionic radius of expression 9.1. Latimer, for example, showed that if 0.85 Å was added to the cation radius and 0.10 Å to the anion radius then the Born prediction that $\Delta G_h \propto r^{-1}$ is held more accurately.

The main impact of the Born equation on the problem of separating the total hydration energy into ionic parts is its implication that there is a correlation between ionic radius and hydration energy. This led Bernal and Fowler to argue that, since K^+ and F^- have almost equal radii, the enthalpy of hydration of KF, $-820 \, kJ \, mol^{-1}$, should divide into equal contributions for the two ions. A similar suggestion has been made that the large ions $AsPh_4^+$ and BPh_4^- would have approximately equal sizes and hence equal hydration enthalpies.

Probably the minimal conclusion that one can draw from the Born equation, and developments of it, is that the hydration enthalpies of ions in the same periodic groups tend monotonically to zero as the ionic radius approaches infinity. In general, the hydration enthalpies that have been deduced from models having their origin in the Born equation satisfy this criterion, as can be seen from Figure 9.2 where some of the 'best' single ion enthalpies collected by Friedman and Krishnan are shown as a function of the Pauling ionic radius.

The trivalent cations that are included in Figure 9.1 do not all fall on a single smooth curve. In the case of the transition metals it is known that there are variations in the hydration enthalpies arising from the ligand-field stabilization of the d electrons. For example, Fe^{3+}, a d^5 system, has a smaller hydration enthalpy than would be expected from a smooth variation within the period. A detailed explanation of this can be found elsewhere.†

It can be seen from Table 9.2 that the enthalpies and free energies of hydration are very similar in magnitude because the entropy of hydration is rather small. The entropies show a correlation with ionic radius which is not too dissimilar from 9.1. Hydration entropies are always negative, which means that the water molecules surrounding an ion have a greater degree of order than they do in the pure liquid.

Gas-phase Hydration

An experimental technique which sheds light on the hydration of individual ions has been developed by Kebarle and co-workers. Ions are produced in the gas phase either by thermionic emission from a filament painted with a

---

†For example, J. N. Murrell, S. F. A. Kettle and J. M. Tedder, *Valence Theory*, 2nd Edn, Wiley, London and New York (1965).

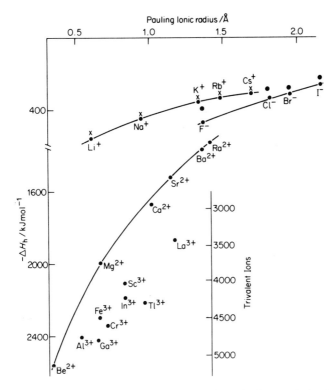

Figure 9.2. Relationships between the enthalpy of hydration and the ionic radius (data from Friedman and Krishnan, *Water, a Comprehensive Treatise*, Vol. 3, Chapter 1, Ed. F. Franks, Plenum Press, London and New York (1973) Table 9.2). Gas-phase enthalpies for the formation of the hexahydrate are shown as separate points: + = alkali metals, ● = halides. Data from Table 9.3

Table 9.2. Enthalpies, entropies and free energies of hydration at 298 K (units kJ mol^{-1} for ΔH_h and ΔG_h and J mol^{-1} K^{-1} for ΔS_h)

	$-\Delta H_h$	$-\Delta S_h$	$-\Delta G_h$		$-\Delta H_h$	$-\Delta S_h$	$-\Delta G_h$
H$^+$	1129	131	1090	Ga^{3+}	4799	577	4628
Li$^+$	559	141	517	In^{3+}	4223	493	4072
Na$^+$	444	110	411	F$^-$	474	133	434
K$^+$	360	74	338	Cl$^-$	340	76	317
Be^{2+}	2563	425	2436	Br$^-$	326	61	303
Mg^{2+}	1998	311	1906	I$^-$	268	38	257
Ca^{2+}	1669	254	1593	OH$^-$	423	149	379
Al^{3+}	4774	530	4616	S^{2-}	1296	86	1270

Values selected by H. L. Friedman and C. V. Krishnan, *Water, a Comprehensive Treatise*, Vol. 3, Ch. 1, Ed. F. Franks, Plenum Press, London and New York (1973).

suitable salt or, for negative ions, by reactive electron transfer with O_2^-. For example, F^- can be produced by

$$O_2^- + NF_3 = O_2 + F^- + NF_2. \tag{9.2}$$

The ions are produced in a reaction chamber containing water vapour and they stay there a sufficient time for equilibrium to be reached in the following general reaction scheme

$$M^\pm(H_2O)_{n-1} + H_2O \rightleftharpoons M^\pm(H_2O)_n. \tag{9.3}$$

The equilibrium concentrations of hydrated ions can be determined by mass spectrometry and, following the principles to be described in Section 10.1, the standard free energy changes can be determined. By varying the temperature in the reaction chamber, the enthalpies and entropies of the reactions can be found (Table 9.3).

Clusters with up to six water molecules per ion have been detected and, if the total enthalpy change for the reaction

$$M^\pm + 6H_2O \rightarrow M^\pm(H_2O)_6 \tag{9.4}$$

is calculated (with extrapolation to $n = 6$ where necessary), then it can be seen from Figure 9.1 that these values show a good correlation with the enthalpies of hydration in solution. From this we can infer that there are

Table 9.3. Enthalpies of the reactions 9.3 $(-\Delta H_m/\text{kJ mol}^{-1})$ for alkali-metal and halide ions

n	Li^+	Na^+	K^+	Rb^+	Cs^+
1	142	100	75	67	57
2	108	83	67	57	52
3	87	66	55	51	47
4	69	58	49	47	44
5	58	51	45	44	—
6	51	45	42	—	—
$\sum_{n=1,6}$	515	403	333	306	282

n	F^-	Cl^-	Br^-	I^-
1	97	55	53	43
2	69	53	51	41
3	57	49	48	39
4	56	46	46	—
5	55	—	—	—
$\sum_{n=1,6}$	389	287	280	228

The summations are the enthalpies for the addition of six water molecules to the ion (extrapolation where necessary). Data from I. Dzidic and P. Kebarle, *J. Phys. Chem.*, **74**, 1466 (1970); M. Ashardi, R. Yamdagni and P. Kebarle, *J. Phys. Chem.*, **74**, 1475 (1970).

probably six water molecules in the first coordination shell around these ions in aqueous solutions.

The hydrated proton has also been studied in the gas phase by Kebarle and co-workers. Table 9.4 gives $\Delta H_h(n)$ the enthalpies of hydration for the reaction

$$H^+(H_2O)_{n-1} + H_2O \rightarrow H^+(H_2O)_n \tag{9.5}$$

and the accumulated enthalpies for the addition of n water molecules. It appears that $-\Delta H_h(n)$ approaches a value of 30–40 kJ mol^{-1} as n approaches infinity, and this is quite close to the enthalpy of evaporation of liquid water which is 44 kJ mol^{-1}. The best estimate of the hydration enthalpy of the proton can be seen from Table 9.2 to be -1129 kJ mol^{-1}, which is broadly consistent with the values given in Table 9.4.

Standard States

The enthalpy change for a chemical reaction is equal to the difference between the sum of the enthalpies of formation of the reactants and the sum of the enthalpies of formation of the products. For pure compounds, enthalpies of formation are by convention referred to a scale in which the enthalpy of every element at its standard state is zero. The standard state is the most stable form of the element at 1 atm and the temperature under consideration; most tabulated values refer to a temperature of 298 K.

For example, if we require the enthalpy change for the reaction

$$CH_4 + 2O_2 \rightarrow CO_2 + 2H_2O(l), \tag{9.6}$$

in which all components are gases except for water, then

$$\Delta H^\ominus = \Delta H_f^\ominus(CO_2) + 2\Delta H_f^\ominus(H_2O, l) - \Delta H_f^\ominus(CH_4) - 2\Delta H_f^\ominus(O_2), \tag{9.7}$$

and from available tables†

Table 9.4. Gas-phase enthalpies (300 K, 1 atm) for the reaction 9.5 and accumulated enthalpies for the addition of n water molecules to H^+

n	1	2	3	4	5	6	7	8
$-\Delta H_h(n)/\text{kJ mol}^{-1}$	690	151	93	71	64	54	49	43
$-\sum_{i=1}^{n}\Delta H_h(i)/\text{kJ mol}^{-1}$	690	841	934	1005	1069	1123	1172	1215

P. Kebarle, S. K. Searles, A. Zolla, J. Scarborough and M. Arshardi. *J. Amer. Chem. Soc.*, **89**, 6393 (1967).

---

†JANAF thermochemical tables. NSRDS technical note, 2nd Edn (1971) and later supplements.

$$\Delta H^{\ominus}(298 \text{ K, 1 atm}) = [-393.5 - 2 \times 285.8 + 74.8 + 0] \text{ kJ mol}^{-1},$$
$$= -890.3 \text{ kJ mol}^{-1}.$$

Free energies of formation are defined with respect to the same zero scale as enthalpies but entropies are by custom always referred to a value of zero for a perfect crystal at absolute zero (the third law of thermodynamics). Thus for the reaction 9.6

$$\Delta S^{\ominus} = S^{\ominus}(CO_2) + 2S^{\ominus}(H_2O, l) - S^{\ominus}(CH_4) - 2S^{\ominus}(O_2) \quad (9.8)$$

and

$$\Delta S^{\ominus}(298 \text{ K, 1 atm}) = [213.6 + 2 \times 69.9 - 186.2 - 2 \times 205.1]$$
$$= -243.0 \text{ J K}^{-1} \text{ mol}^{-1}$$

For the sole objective of calculating entropy changes for a reaction, it would have been possible to have adopted the same procedure as for enthalpy and free energy, and taken as zero the entropies of formation of the elements in their standard states.

Since ions of one type are always formed in association with counter ions, it is impossible to determine the absolute values of the thermodynamic parameters of a single ionic species. Of course, absolute values of energies are never available, and it is a familiar exercise to refer energies to a convenient arbitrary standard, as mentioned on the previous page. In this case, all that is necessary is to assign arbitrary values to $\Delta H_f^{\ominus}$, $\Delta G_f^{\ominus}$ and $S^{\ominus}$ for one ion in order to be able to tabulate values of these quantities for all other ions; by convention, these three parameters are all assigned the value zero for the hydrated proton (H^+, aq). Note that entropies so derived are still indicated as absolute quantities although they have no connection with the third-law scale. Third-law entropies for pure substances are always positive, but $S^{\ominus}$ values for ions in solution may be positive or negative.

The enthalpy of solution of HCl gas is -75 kJ mol^{-1} and the enthalpy of formation of HCl gas is -92 kJ mol^{-1}. We therefore have for the process

$$HCl(g) \rightarrow H^+(aq) + Cl^-(aq), \quad (9.9)$$
$$-75 \text{ kJ mol}^{-1} = \Delta H_f^{\ominus}(H^+, aq) + \Delta H_f^{\ominus}(Cl^-, aq) - \Delta H_f^{\ominus}(HCl, g)$$
$$= [0 + \Delta H_f^{\ominus}(Cl^-, aq) + 92] \text{ kJ mol}^{-1}$$

whence

$$\Delta H_f^{\ominus}(Cl^- \text{ aq}) = -167 \text{ kJ mol}^{-1}.$$

By successive consideration of such reactions, a $\Delta H_f^{\ominus}$ table can be constructed for all ions based upon the standard $\Delta H_f^{\ominus}(H^+, aq) = 0$. It is important to note that the relative values for cations and anions in this scale have no direct physical significance. If $\Delta H_f^{\ominus}(H^+, aq)$ had been given some other value, A (say) then $\Delta H_f^{\ominus}(Cl^-, aq)$ would have been reduced by A, so that the *difference* between the $\Delta H_f^\ominus$ values for the two ions would become $2A$.

There is an inconsistency in the convention that for the reaction

$$\tfrac{1}{2}H_2(g) \rightarrow H^+(aq), \qquad (9.10)$$

$\Delta G^\ominus$ and $\Delta H^\ominus$ are zero but $\Delta S^\ominus$ is not zero (because $S^\ominus$ for H_2 is not zero). However, this inconsistency has no practical effect for reactions in which a charge balance is retained. For example, for

$$\tfrac{1}{2}H_2(g) + \tfrac{1}{2}Cl_2(g) \rightarrow H^+(aq) + Cl^-(aq)$$

the values of $\Delta G^\ominus$, $\Delta H^\ominus$ and $\Delta S^\ominus$ calculated from standard tables, satisfy the requirement that $\Delta G = \Delta H - T\Delta S$.

9.2 IONIC TRANSPORT AND HYDRATION NUMBERS

Much of our present knowledge of the nature of electrolyte solutions stems from the study of their electrical conductivity by such scientists as Faraday, Kohlrausch and Arrhenius in the nineteenth century. Kohlrausch, for example, showed that the conductivity per mole of electrolyte, when extrapolated to infinite dilution, could be expressed as the sum of contributions from individual ions. He published his results in 1855 as the law of independent migration of ions.

The resistance R of a material is proportional to its length l and inversely proportional to its cross-section area A. The proportionality factor is called the resistivity, ρ,

$$R = \rho l / A \qquad (9.11)$$

and the inverse of ρ is the conductivity κ. The conductivity of pure water is very low, $\kappa \approx 1 \times 10^{-8}\,\Omega^{-1}\,cm^{-1}$ ($\Omega = $ ohm) at $273\,K$ so that, even for electrolyte solutions of low concentration, all the conductivity can be attributed to the ions, and for strong electrolytes the conductivity is roughly proportional to the concentration of the salt.

As a measure of the conductivity for unit concentration of salt, we use the molar conductivity defined by

$$\Lambda_m = \kappa / c \qquad (9.12)$$

where c is the molarity. Because ions carry different charges a better measure of the speed at which ions travel through the solution is given by the equivalent conductance. This is obtained from Λ_m by dividing by the number of Faradays carried by a mole of salt; the divisor is one for NaCl, two for Na_2SO_4, six for $Fe_2(SO_4)_3$, etc.

Figure 9.3 shows the molar conductivities of some strong electrolytes in dilute solutions. As mentioned earlier, these conductivities are not rapidly

198

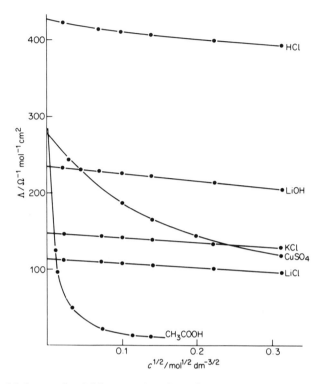

Figure 9.3. Molar conductivities as a function of concentration for some electrolyte solutions at 298 K. (Data from C. W. Davies, *Electrochemistry*, Newnes, London (1967))

varying functions of concentration, but what variation there is follows the relationship

$$\Lambda_m = \Lambda_m^0 + ac^{1/2} \tag{9.13}$$

where Λ_m^0 is the conductivity at infinite dilution. This empirical relationship was first noted by Kohlrausch. We note in addition that all the 1:1 electrolytes have approximately the same value of a. There is in fact a theory of ion transport by Onsager (more advanced than the level we reach in this book) which shows that a has the form $A + B\Lambda_m^0$, where both A and B depend on the charges carried by the ions as well as on bulk properties of the solution such as relative permittivity and viscosity.†

Kohlrausch noted that the difference in Λ_m^0 values for pairs of salts having a common ion were independent of that ion, as is illustrated by the values given in Table 9.5. This behaviour can be attributed to the independent mobility of the two ions, and stimulated Kohlrausch to write

†H. S. Harned and B. B. Owen, *The Physical Chemistry of Electrolyte Solutions*, Reinhold. New York (1958).

Table 9.5. Values of Λ_m^0 for some electrolyte solutions in water at 298 K and their differences

	System	Λ_m^0	2–1 4–3	3–1 4–2
1	LiCl	115		
2	NaCl	126	11	122
3	LiOH	237		122
4	NaOH	248	11	

Units: $\Omega^{-1}\,mol^{-1}\,cm^2$.

$$\Lambda_m^0 = \Lambda_m^{0+} + \Lambda_m^{0-} \qquad (9.14)$$

where $\Lambda_m^{0\pm}$ are the molar ionic conductances at infinite dilution.

We have already seen that to write some property as the sum of separate ion contributions does not necessarily mean that there is an unambiguous method of obtaining the separate ion values; hydration energy is a case where the separation can only be made with certain assumptions. However, for ionic mobilities it is possible to determine the separate ion values directly by experiment. The technique involves measuring changes in the concentration of the ions arising from the passage of an electric current through the solution.

The fraction of the current carried by an ion in solution is called its transport or transference number, t_0^+ or t_0^-. These are related to the ionic conductance by the expressions

$$t_0^+ = \frac{\Lambda_m^{0+}}{\Lambda_m^{0+} + \Lambda_m^{0-}}, \qquad t_0^- = \frac{\Lambda_m^{0-}}{\Lambda_m^{0+} + \Lambda_m^{0-}}. \qquad (9.15)$$

It should be noted that these transport numbers are not parameters characteristic of the individual ions but of the electrolyte as a whole, because both ion conductances are involved in their definitions.

If v^+ and v^- are the velocities of the two ions in solution in the direction of the applied field then

$$\frac{t_0^+}{t_0^-} = \frac{|Z^+ v^+|}{|Z^- v^-|} \qquad (9.16)$$

where Z^+ and Z^- are the ion charges. It follows that if we measure the ratio v^+/v^- we also have a measure of $\Lambda_m^{0+}/\Lambda_m^{0-}$. One method of measuring this ratio was developed by Hittorf, and the principles of the method are shown in Figure 9.4.

To carry out the Hittorf experiment one requires an electrochemical cell with three compartments: an anode compartment, a cathode compartment and a central connecting compartment. It is important that there is no mixing of solutions between these three compartments by convection, so

200

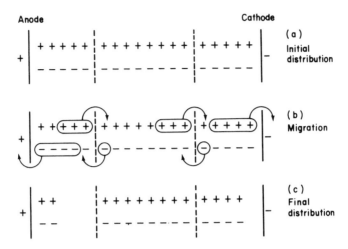

Figure 9.4. The changes in concentration of electrolyte near the electrodes which arise from different transport numbers of the ions (Hittorf experiment)

that changes in concentration are entirely due to the migration of ions under the influence of the electric field.

If the cations are three times as mobile as the anions then it can be seen from Figure 9.4 that in the anode compartment three times more cations will leave in a given time than anions enter. In the cathode compartment three times more cations enter than anions leave. The potential excess charge in each compartment which would be produced by such movements is neutralized by discharge of the ions at the electrode, and it can be seen (Figure 9.4(c)) that this produces a build-up of electrolyte concentration in the cathode compartment. The ratio of the changes in concentration of electrolyte, $\Delta n_c/\Delta n_a$, in the cathode and anode compartments is equal to the ratio of the velocities of anion and cation, and hence, by 9.15 and 9.16, the ratios of the ionic conductances can be obtained. In this example it is necessary to discharge four equivalents of the ions to maintain electrical neutrality in the anode and cathode compartments.

Other techniques for measuring transport numbers are described in more comprehensive texts on electrolyte solutions.†

Table 9.6 gives some values of ionic conductances in water at 298 K and shows the temperature dependence for Na^+ and Cl^-. One of the most interesting points concerning these values is that they do not show the expected correlation with ionic radius. The radius increases down a group in the periodic table (see Figure 9.2) yet the conductances, and hence mobilities, of both the alkali metal and halide ions also increase down the group. The reason for this behaviour is that the species being carried along

---

†H. S. Harned and B. B. Owen, see reference on p. 198.

Table 9.6. Ionic conductances at infinite dilution in water at 298 K $\Lambda_m^{0\pm}/\Omega^{-1}\,mol^{-1}\,cm^2$) and for Na^+ and Cl^- at other temperatures

	Λ_m^{0+}		Λ_m^{0-}		$\Lambda_m^{0\pm}$
H^+	350	OH^-	192	NH_4^+	40
Li^+	39	F^-	55	Me_4N^+	44
Na^+	51	Cl^-	76	nBu_4N^+	19
K^+	74	Br^-	78	NO_3^-	71
Ca^{2+}	120	I^-	77	SO_4^{2-}	158
Ba^{2+}	130				
T/K	273	298	323	348	373
Na^+	26	51	82	116	155
Cl^-	41	76	116	160	207

The equivalent conductances of doubly charged species are half the above values. Data from *Handbook of Chemistry and Physics*, 59th Edn, Chem. Rubber Publ. Co. (1979) and R. L. Kay and D. F. Evans, *J. Phys. Chem.*, **70**, 2325 (1966).

in an electric field is not the bare ion but the hydrated ion. Because the hydration energies decrease from Li^+ to Cs^+ and from F^- to I^-, so the effective size of the hydrated ion also decreases.

The effect of temperature on the ionic conductance can also be explained by hydration. As the temperature is increased so the average number of water molecules bound to the ion decreases. However, although the ions carry a hydration shell with them as they move through the solution, it must not be thought that this shell is permanently attached to the ion. A more realistic picture is that the hydration shell continuously exchanges water molecules as the ion passes through the solution, some dropping off and other taking their place. We can in fact place some limits on the average time a water molecule spends within a hydration shell by nuclear magnetic resonance spectroscopy.

N.m.r. is by the time-scale of molecular motion rather a slow experiment. If a magnetic nucleus has two environments whose n.m.r. frequencies, v_1 and v_2, differ by Δv, then these will appear as separate lines in the spectrum only if the average time for exchange between the sites is appreciably greater than Δv^{-1}. If the exchange time is appreciably less than Δv^{-1}, the system will show only a single line in the spectrum whose frequency is a population average of v_1 and v_2.

Typical frequency differences (chemical shifts) in proton n.m.r. are 10^2–10^3 Hz so that exchange times faster than 10^{-3}–10^{-2} s will lead to frequency averaging. Other nuclei can give a much bigger spread of chemical shift. For water O^{17} has proved particularly useful: in the presence of paramagnetic ions, chemical shift differences of 10^5 Hz can be obtained between water molecules in a solvation shell and water in the bulk.

At room temperature, aqueous solutions of electrolytes containing only diamagnetic ions show frequency-averaged proton n.m.r. spectra but on cooling to around 223 K separate frequencies for the solvation shell and free molecules can be seen for many multicharged cations. For Al^{3+}, for example, two frequencies separated by approximately 270 Hz have been observed† showing that at these temperatures the exchange time between water in the solvation shell and in the bulk is greater than about 10^{-3} s. However, for the alkali-metal ions and for all anions, only exchange-averaged spectra are observed even at low temperatures.

Because the intensity of a line in an n.m.r. spectrum is proportional to the number of nuclei which contribute to that line, it is possible by measuring the relative intensities of the two lines in the non-exchanged spectra to deduce the relative number of solvation shell and bulk water molecules. In the case of Al^{3+} solutions it was found from this technique that there were six water molecules in the hydration shell. This number has been found to be remarkably uniform for the cations studied by n.m.r. With pure water as a solvent only Be^{2+} has been found to have a different hydration number, approximately four, although other low numbers have been found for mixed solvent systems.

Several review articles‡ provide a more detailed discussion of the application of n.m.r. to solvation studies. To summarize the most important result: for multicharged cations at room temperature, the exchange of water molecules between the solvation shell and the bulk takes about 10^{-5} s, and for monovalent cations and anions it is considerably faster.

The concept of hydration number has often been abused in the past. If it is taken to mean the number of water molecules around an ion whose properties can be distinguished from those of water in the bulk, then the number depends on what property is being measured. Hydration numbers of the order of 20 have been derived from the analysis of some experiments. N.m.r. undoubtedly indicates the number of water molecules which are directly in contact with the ion. If there are second shells of molecules which are significantly influenced by the electrostatic field of the ion, then they exchange too quickly with bulk water to be studied by n.m.r.

X-ray and neutron scattering from aqueous electrolyte solutions confirm the n.m.r. result that six is the most common hydration number for monatomic ions, and give some structural information about the hydration shell. Of the two techniques, neutron diffraction appears to have more promise for the future because by using isotopic substitution some of the complicating features such as incoherent scattering and inelastic scattering (energy transfer in the scattering process) can be eliminated.

Figure 9.5 shows a radial distribution function $\overline{G(r)}$ that has been obtained by taking the Fourier transform of the difference in scattering

†R. E. Schuster and A. Fratiello, *J. Chem. Phys.*, **47**, 1554 (1967).

‡For example: A. Fratiello, *Prog. Inorg, Chem.*, **17**, 57 (1972).

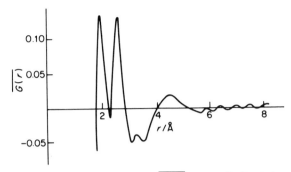

Figure 9.5. The radial correlation function $\overline{G(r)}$ for $NiCl_2/D_2O$ based on isotopic substitution of Ni. (J. E. Enderby and G. W. Neilson, *Water, a Comprehensive Treatise*, Ed. F. Franks, Vol. 7, Chapter 1, Plenum Press, London and New York (1979)

factors for naturally occurring $NiCl_2$ in D_2O and for $^{62}NiCl_2/D_2O$. For details of the analysis the reader is referred to the original papers;† it is sufficient for our purposes to think of $\overline{G(r)}$ as a function like $g(r)$ (equation 1.13) which gives the radial correlation between Ni^{2+} and surrounding nuclei.

The two peaks in Figure 9.5 at 2.07 Å and 2.67 Å are identified with Ni–O and Ni–D correlations, respectively. The areas beneath them when weighted by r^2 are almost exactly in the ratio 1:2 as required. The integral over $4\pi r^2 \overline{G(r)}$ between 1.8 and 3.0 Å yields a total of 5.8 ± 0.2 water molecules in the first coordination shell. The low value of $\overline{G(r)}$ between 3 and 4 Å suggests that the first hydration shell is very tight.

If it is assumed that the geometry of the water molecules in the solvation shell is the same as in the bulk then from the Ni–O and Ni–D peak distances it is concluded that the two-fold symmetry axis of water makes an angle of approximately 40° to the Ni–O line, as shown in Figure 9.6.

Similar studies of $CaCl_2$ and NaCl solutions using chlorine isotopes show that the Cl^- ions also have approximately six water molecules in the hydration shell but in this case one Cl–D distance is less than the Cl–O distance. Figure 9.6 shows the anion–water structure that has been deduced from these data.

It can be seen from Figures 9.5 and 9.6 that neutron diffraction is giving very interesting information about the structure of electrolyte solutions. This now extends to the nature of ion–ion distribution functions, a topic which we cover in Section 9.3.

The structures of the hydrated proton and the hydroxyl anion are very important for understanding the properties of aqueous solutions of acids and bases. It can be seen from Table 9.2 that the enthalpy of hydration of the

†For a review see J. E. Enderby and G. W. Neilson, *Water, a Comprehensive Treatise*, Ed. F. Franks, Vol. 7, 1, Plenum Press, London and New York (1979).

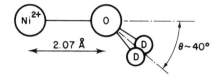

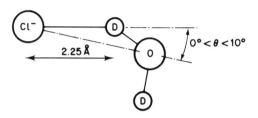

Figure 9.6. Geometries of the solvating water molecules around Ni^{2+} and Cl^- that have been deduced by neutron diffraction. The Ni^{2+}–H_2O geometry applies to solutions more concentrated than 1.46 molar. In more dilute solutions θ decreases and possibly becomes zero

proton is much more negative than that of any other singly-charged cation, and from Table 9.4 that much of this stabilization comes by adding the first water molecule. The convention of writing the proton in water as the species H_3O^+ is consistent with these values, but gives no indication of how the H_3O^+ ion is hydrated.

Quantum mechanical calculations on H_3O^+ show that is is pyramidal with bond angles of about 114°: neutron diffraction studies of some solid acid hydrates give a value a little less than this, approximately 110°. As the hydrogen atoms in H_3O^+ carry a substantial net positive charge they would be expected to form three strong hydrogen bonds to other water molecules in aqueous solution. Calculations confirm this and predict the equilibrium structure for $H_9O_4^+$ shown in Figure 9.7.† This is a structure first postulated by Eigen as the best representation of the hydrated proton.

An interesting question is whether the oxygen atom in H_3O^+ can be involved in hydrogen bonding. Calculations show that if another water molecule is added to $H_9O_4^+$ it adds preferentially to one of the external H_2O fragments rather than to the central H_3O^+.

In contrast to H_3O^+ the species OH^- forms hydrogen bonds only through its oxygen atom. Three strong bonds are formed with water molecules to give the species $H_7O_4^-$ whose structure is shown in Figure 9.7.

From Table 9.6 it can be seen that the molar conductivities of H^- and OH^- are much larger than those of the other ions listed. From their

†M. D. Newton and S. Ehrenson, *J. Amer. Chem. Soc.*, **93**, 4971 (1971); M. D. Newton, *J. Chem. Phys.*, **67**, 5535 (1977).

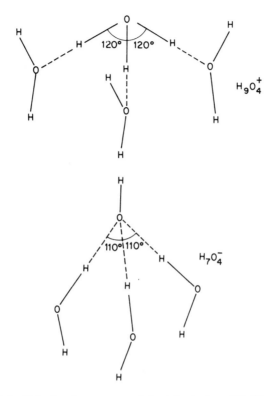

Figure 9.7. Calculated structures of the trihydrates of H_3O^+ and OH^-

enthalpies of hydration we know that they have a tightly bound solvent shell; one might expect H^+ to be less mobile than Li^+ and OH^- to have a mobility comparable to that of F^-. As no rationalization of the data can be made on the basis of ion size, or hydration enthalpy, we deduce that the mechanism of conduction for these ions in water is different from that of other ions.

Figure 9.8 illustrates schematically a special mechanism of conduction for the proton (there is not, of course, such perfect alignment in real water). A proton from OH_3^+ passes to one of the H_2O molecules to which it is hydrogen bonded and this molecule then passes on one of its other protons to a neighbour: the proton received is not the proton passed on. The mechanism for OH^- mobility is similar; in this case a proton is transferred from H_2O to OH^- and another proton comes in to replace it along a different hydrogen bond direction.

The question remains as to why the molar conductivity of OH^- is only about half the value for H^+ when the mechanisms for conduction appear to be so similar. There is evidence from the rates of acid-base reactions that the transfer of protons between molecular species depends to a considerable extent on quantum mechanical tunnelling through potential energy

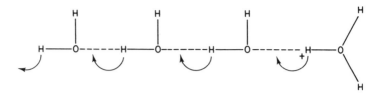

Figure 9.8. The mechanism of proton conduction in water

barriers.† This is evident from the deuterium isotope effect on such rates. We therefore expect the proton migration involved in H^+ and OH^- transport to be sensitive to small details of the hydrogen bonding of water to H_3O^+ and OH^-. The binding energy of three water molecules to H_3O^+ in $H_9O_4^+$ is slightly greater than to OH^- in $H_7O_4^+$ and we infer that such differences have a large effect on the proton mobility in the two cases.

9.3 THE IONIC ATMOSPHERE

Electrolyte solutions depart strongly from ideality. The evidence for this is extensive: from colligative properties such as the depression of freezing point; from the molar conductivity, as shown in Figure 9.3; and from many other sources.

For many years this non-ideal behaviour was attributed to ion pairing and, following the Arrhenius treatment of weak electrolytes, the concentration of ions in solution was calculated by assuming that an equilibrium existed between free ions and undissociated salt molecules. However, the degree of dissociation that had to be assumed to account for, say, the molar conductivity, was in conflict with other evidence, particularly spectroscopic, that the dissociation was almost complete.

By 1910 it was becoming clear that strong electrolytes were completely dissociated in solution, and that their non-ideality was due to the Coulomb interaction between ions extending over large distances. In 1912 Milner proposed a mathematical formulation of the problem which was subsequently shown to be essentially correct, but it was not until 1923 that a satisfactory solution of the equations describing this long-range interaction was produced by Debye and Hückel. The delay is remarkable in view of the fact that in 1910 Gouy and Chapman produced a mathematically similar theory of the distribution of ions outside a charged electrode (Chapter 13).

Although there are contentious steps in the development of the Debye–Hückel theory, it has proved to be very successful in predicting the non-ideal behaviour of very dilute solutions of strong electrolytes. The mathematical details of their solution are, we feel, less important than the underlying

---

†See R. P. Bell, *The Proton in Chemistry*, Ch. 11, Methuen, London (1959).

physical principles; for that reason we concentrate here on these principles and leave most of the mathematics to an appendix at the end of this chapter.

Because like charges repel and unlike charges attract one another, there will, in electrolyte solutions, be a preference for an anion to have cations in its vicinity and for a cation to have anions in its vicinity. This situation also occurs in ionic solids where an ion is surrounded by alternating shells of anions and cations at discrete radial distances as illustrated in Figure 9.9(a) for the NaCl crystal. However, in dilute solutions we expect the average charge distribution arising from the surrounding ions to be a smoothly decaying function of r as shown qualitatively in Figure 9.9(b). An unresolved question is whether the charge density function in concentrated electrolyte solutions has oscillatory features suggestive of the solid state.

The time-averaged distribution of charge surrounding an ion is called its ionic atmosphere. The Debye–Hückel theory is concerned with finding the functional form of this atmosphere and deducing its effect on the chemical potentials of the ions.

The electrostatic potential ϕ at distance r from a single ion of charge Ze in a solvent of relative permittivity ε_r is

$$\phi = \frac{Ze}{4\pi\varepsilon_0\varepsilon_r}\left(\frac{1}{r}\right). \tag{9.17}$$

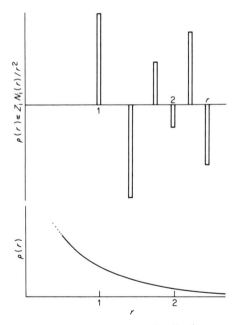

Figure 9.9. (a) A representation of the charge distribution around an anion in the NaCl crystal. The ions are treated as point charges; N_i ions of charge $Z_i e$ at distance r (measured in lattice units). (b) A representation of the charge distribution around an anion in dilute solution. Arbitrary units for $\rho(r)$ in both figures

The ionic atmosphere surrounding this ion will partly shield this Coulomb potential at short distances. To preserve electrical neutrality for the system as a whole, the integral of the electrical charge in the ionic atmosphere from $r = 0$ to infinity must be equal and opposite to the charge on the ion. It follows that at a large distance the ionic atmosphere will completely screen the potential of the ion. We can therefore write the potential of the ion and its atmosphere as

$$\phi = \frac{Ze}{4\pi\varepsilon_0\varepsilon_r}\left(\frac{F(r)}{r}\right), \tag{9.18}$$

where

$$\underset{r\to\infty}{\text{Limit}}\, F(r) = 0. \tag{9.19}$$

It will be shown in Appendix 9 that the Debye–Hückel theory in its simplest form leads to the result

$$F(r) = \exp(-r/r_D), \tag{9.20}$$

where r_D, called the Debye length, can be thought of as the effective radius of the ionic atmosphere.

We shall see that r_D depends on various parameters of the solution and that in qualitative terms the nature of this dependence can be deduced by physical arguments. For example, an increase in temperature will tend to break up any ordered structure in the solution, hence we expect r_D also to increase. Similarly the dielectric effect reduces the electrostatic energy between ions so we expect r_D to increase with increasing ε_r.

The most important factors determining r_D are the charges and concentrations of the ions. The Coulomb interaction between two ions is proportional to the product of their charges, hence we expect r_D to decrease if the charges are increased. With increasing concentration the ions will, on average, be closer together and hence r_D will again decrease.

The relationship between the non-ideality of electrolyte solutions and the charge and concentration of ions was first discovered empirically by Lewis and Randall. To describe their findings they defined a quantity called the ionic strength I as

$$I = \frac{1}{2}\sum_i m_i Z_i^2, \tag{9.21}$$

where m_i is the molality of the ion of charge $Z_i e$, and they arrived at the following conclusion.[†] 'Aside from the specific characters of individual ions there is unquestionably a factor depending upon the concentration of the ions and upon their charge, which largely determines the properties of any electrolyte solution; and we feel confident that this factor is the ionic

---

[†] G. N. Lewis and M. Randall, *Thermodynamics*, McGraw-Hill, Toronto and New York (1923).

strength.' Their confidence has proved justified, for the ionic strength is the most important factor controlling the deviation from ideality of dilute electrolyte solutions.

It can be seen from the definition 9.21 that the ionic strength is proportional to the electrolyte concentration, but the proportionality factor depends on the ion charges. For a 1:1 electrolyte like NaCl, $I = m$. For a 1:2 electrolyte like Na_2SO_4 of molality m we have

$$m(Na^+) = 2m, \ m(SO_4^{2-}) = m,$$

and hence

$$I = \tfrac{1}{2}(2m.1^2 + m.2^2) = 3m. \tag{9.22}$$

For a 2:2 electrolyte, $I = 4m$, and so on.

We show in the appendix to this chapter that the simplest solution of the equations governing the ionic atmosphere gives the following expression for the Debye length (equation A9.12)

$$r_D = \left[\frac{\varepsilon_0 \varepsilon_r RT}{2d.\mathcal{N}^2 e^2 I} \right]^{1/2}. \tag{9.23}$$

where d is the density of the solvent in $kg\,m^{-3}$. This expression satisfies the qualitative statements made above regarding the relationship between r_D and T, ε_r and I. For water at 298 K

$$r_D = 3.04 I^{-1/2} \text{ Å}, \tag{9.24}$$

where I is measured in $mol\,kg^{-1}$.

To derive 9.23 it has to be assumed that the solvent can be treated as a continuous dielectric and that the ions can be approximated as point charges. These assumptions can only be valid for solutions which are sufficiently dilute that the average separation of the ions is large compared with the size of solvent molecules and the ions themselves. If r_D is only a few ångströms the assumptions are clearly invalid because the ions would overlap each other. If we take a rather generous limit of validity of $r_D \approx 10$ Å this gives an upper limit to the concentration of a 1:1 electrolyte in water at 298 K of $0.1\,mol\,kg^{-1}$, and for a 2:2 electrolyte of only $0.025\,mol\,kg^{-1}$. Thus, the theory can only be expected to predict accurately the behaviour of quite dilute solutions.

9.4 THE ACTIVITIES OF IONS IN SOLUTION

We follow the practice already established for non-electrolyte solutions (6.67) and refer the chemical potential to a convenient standard state using an expression of the form

$$\mu = \mu^\ominus + RT \ln a, \tag{9.25}$$

where a is the activity. Ideal behaviour of the solute in a non-electrolyte solution was defined as a system obeying Henry's law for which we could write (6.73)

$$\mu = \mu^{\ominus} + RT \ln x. \tag{9.26}$$

For electrolytes it is more common to express solute concentrations as molality, so that ideal behaviour can be represented by

$$\mu = \mu^{\ominus} + RT \ln (m/\text{mol kg}^{-1}). \tag{9.27}$$

Note, from 6.4, that in dilute solutions x is proportional to m, so that 9.27 is equivalent to 9.26 except for a redefinition of $\mu^{\ominus}$; this is now the chemical potential of the solute in an ideal system at unit molality. Moreover, as electrolytes are usually involatile, the connection between ideality and Henry's law is now purely formal.

For the chemical potential of the solute in real solutions we therefore adopt expression 9.25 in which the activity is a dimensionless quantity that approaches the numerical value of m as m approaches zero,

$$\underset{m \to 0}{\text{Limit}}\, a = m/\text{mol kg}^{-1}. \tag{9.28}$$

Further, if we introduce an activity coefficient γ (also dimensionless) as in 6.65 by

$$a = \gamma m/\text{mol kg}^{-1}, \tag{9.29}$$

then

$$\underset{m \to 0}{\text{Limit}}\, \gamma = 1. \tag{9.30}$$

The chemical potential attains its standard-state value when $a = 1$. If the system were ideal we could write $a = m = 1$ but for real systems this is not true, indeed when $a = 1$ the solutions are far from ideal. Following the same policy as for the perfect gas and for the solute in non-electrolyte solutions, we therefore note that the standard state is not a real state that can be achieved for the system. The value of $\mu^{\ominus}$ assigned to it can, however, be obtained by extrapolating the value of $\mu - RT \ln m$ to the ideal limit of $m = 0$.

If we adopt expression 9.25 for each type of ion in solute ion then the chemical potential for a strong electrolyte that ionizes as follows

$$M_p X_q \rightleftharpoons pM^{q+} + qX^{p-}, \tag{9.31}$$

is given by

$$\mu = p\mu_+ + q\mu_- = p\mu_+^{\ominus} + q\mu_-^{\ominus} + RT \ln a_+^p a_-^q,$$
$$= p\mu_+^{\ominus} + q\mu_-^{\ominus} + RT \ln m_+^p m_-^q + RT \ln \gamma_+^p \gamma_-^q. \tag{9.32}$$

However, we have seen in Section 9.1 that there is no unique way of separating the thermodynamic properties of electrolytes into separate ion values and hence no way of deducing separate values for γ_+ and γ_-. Moreover, the deviation from ideality arises from the electrostatic interaction between ions and it is not at all obvious that separate and different ion activity coefficients would have any meaning. For this reason, only the average ionic activity coefficient is a measurable quantity and it is defined as the geometric mean of γ_+ and γ_- as follows:

$$\gamma_\pm = (\gamma_+^p \gamma_-^q)^{1/i}, \tag{9.33}$$

where $i = p + q$ is the number of ions produced by one molecule of the electrolyte. For a 1:1 electrolyte $\gamma_\pm = (\gamma_+ \gamma_-)^{1/2}$.

Introducing 9.33 into 9.32 gives

$$\mu = p\mu_+^\ominus + q\mu_-^\ominus + RT \ln m_+^p m_-^q + iRT \ln \gamma_\pm, \tag{9.34}$$

and the separate ion chemical potentials are defined by

$$\mu_+ = \mu_+ + RT \ln m_+ + RT \ln \gamma_\pm, \tag{9.35}$$

$$\mu_- = \mu_-^\ominus + RT \ln m_- + RT \ln \gamma_\pm. \tag{9.36}$$

The strategy adopted by Debye and Hückel to relate $\gamma_\pm$ to the Debye length r_D was to assume that *all* the non-ideal behaviour of strong electrolyte solutions arises from the electrostatic interaction between the ions. Although solvents which are non-electrolytes exhibit some non-ideal behaviour the departure from ideality is generally much smaller than for electrolytes so that this assumption may reasonably be expected to be valid.

The model for an ideal solution is one in which the ions *have the same distribution as in the real solution* but in which the electrostatic interaction between an ion and its atmosphere is ignored (but not the ion-solvent interaction). The chemical potential of an ion under these conditions, μ^*, is

$$\mu^* = \mu^\ominus + RT \ln m, \tag{9.37}$$

with a $+$ or $-$ suffix, as appropriate. Taking the difference between this and 9.35 (or 9.36) gives

$$\mu - \mu^* = RT \ln \gamma_\pm. \tag{9.38}$$

The difference $\mu - \mu^*$ can now be equated to the change in chemical potential, or work done, on raising the ion charge from zero to its final value Ze in the potential produced by its ionic atmosphere.

Expression 9.18 gives the potential due to the ion and its atmosphere, hence the potential due to the atmosphere alone is

$$\phi'(r) = \frac{Ze}{4\pi\varepsilon_0\varepsilon_r r}(e^{-r/r_D} - 1) \tag{9.39}$$

We can obtain an approximate value of this quantity at $r = 0$ by expanding the exponential and taking the leading term; this gives the result

$$\phi'(0) = -\frac{Ze}{4\pi\varepsilon_0\varepsilon_r r_D}. \tag{9.40}$$

On building up the charge of the ion from zero to Ze the work done per mole is†

$$W = \mathcal{N} \int_0^{Ze} \phi'(0) \, \mathrm{d}(Ze) = -\frac{Z^2 e^2 \mathcal{N}}{8\pi\varepsilon_0\varepsilon_r r_D}, \tag{9.41}$$

and equating this to the change in the chemical potential we have

$$\ln \gamma = \frac{-Z^2 e^2}{8\pi\varepsilon_0\varepsilon_r RTr_D}. \tag{9.42}$$

There will be an expression like 9.42 for each ion in solution. For an M_pX_q electrolyte (9.31) we have, from 9.33

$$\ln \gamma_\pm = i^{-1}(p \ln \gamma_+ + q \ln \gamma_-), \tag{9.43}$$

and introducing the appropriate expression from 9.42

$$\ln \gamma_\pm = -i^{-1}(pZ_+^2 + qZ_-^2)\frac{e^2 \mathcal{N}}{8\pi\varepsilon_0\varepsilon_r RTr_D}. \tag{9.44}$$

This can be simplified by noting that for electrical neutrality

$$pZ_+ + qZ_- = 0, \tag{9.45}$$

whence

$$pZ_+^2 + qZ_-^2 + (p + q)Z_+Z_- = 0, \tag{9.46}$$

and

$$i^{-1}(pZ_+^2 + qZ_-^2) = -Z_+Z_-. \tag{9.47}$$

It follows that expression 9.44 becomes

$$\ln \gamma_\pm = -|Z_+Z_-|\frac{e^2 \mathcal{N}}{8\pi\varepsilon_0\varepsilon_r RTr_D}, \tag{9.48}$$

where, as is customary, the positive value of the product Z_+Z_- has been introduced. If we substitute expression 9.23 for r_D we have

$$\ln \gamma_\pm = -\frac{|Z_+Z_-|e^3\mathcal{N}^2(2dI)^{1/2}}{8\pi(\varepsilon_0\varepsilon_r RT)^{3/2}} \tag{9.49}$$

which is known as the Debye–Hückel limiting law (limiting meaning applicable to dilute solutions). A working expression for water at 298 K is

$$\ln \gamma_\pm = -1.172|Z_+Z_-|(I/\text{mol kg}^{-1})^{1/2}. \tag{9.50}$$

The chemical potentials or activities of strong electrolytes in solution can be determined by a wide variety of techniques. One can establish an equilibrium between the solution of interest and another phase which either

---

†As r_D is a function of the ionic charge it might be thought that this factor should be allowed as a variable when evaluating the integral. This is not so because the work done is being calculated for charging the ion in an ionic atmosphere whose Debye length is fixed.

can be treated as ideal (e.g. the dilute gas or solid), or is one whose chemical potentials have previously been determined. Most electrolytes are involatile and hence measuring the solute vapour pressure is impractical.

The indirect methods, already described for non-electrolytes, require the determination of solvent activites either by vapour pressure measurements or from a colligative property, followed by application of the Gibbs–Duhem equation. Freezing point measurements have extensively been used in this way although care is required in both measurement and analysis to obtain accurate results.

The expression relating the solute activity to the freezing point depression ΔT_m was given in 6.82. The only change we make to it is to use the molality of the solute as the variable instead of the mole fraction. The two are related by 6.3, and we note in particular

$$\frac{1-x}{x} = \frac{1000}{mM}, \tag{9.51}$$

thus expression 6.82 becomes

$$\ln\left(a(m)/a(m^\circ)\right) = \frac{1000\Delta H_{\text{melt}}}{RT_m^2 M} \int_{m^\circ}^{m} \frac{1}{m}\left(\frac{\mathrm{d}\Delta T_m}{\mathrm{d}m}\right)\mathrm{d}m,$$

$$= \frac{1}{K_t} \int_{m^\circ}^{m} \frac{1}{m}\left(\frac{\mathrm{d}\Delta T_m}{\mathrm{d}m}\right)\mathrm{d}m, \tag{9.52}$$

where K_t is the cryoscopic constant (6.59).

For an $M_p X_q$ electrolyte we have from 9.32 and 9.33

$$a = a_+^p a_-^q = (m\gamma_\pm)^i, \tag{9.53}$$

whence

$$\ln a = i(\ln m + \ln \gamma_\pm), \tag{9.54}$$

expression 9.52 therefore becomes

$$\ln\left(m/m^\circ\right) + \ln\left(\gamma_\pm/\gamma_\pm^\circ\right) = \frac{1}{iK_t} \int_{m^\circ}^{m} \frac{1}{m}\left(\frac{\mathrm{d}\Delta T_m}{\mathrm{d}m}\right)\mathrm{d}m. \tag{9.55}$$

If we let the lower limit of the integral m° approach zero then $\gamma_\pm^\circ$ approaches unity. However, there are difficulties in evaluating the integral by graphical means because $m^{-1} \to \infty$ and on the left-hand side of 9.55 $\ln m^\circ \to -\infty$. The solution is best obtained by introducing a new variable

$$j = 1 - \frac{\Delta T_m}{iK_t m}, \tag{9.56}$$

from which we have

$$\mathrm{d}j = (1 - j)\frac{\mathrm{d}m}{m} - \frac{1}{iK_t m}\left(\frac{\mathrm{d}\Delta T_m}{\mathrm{d}m}\right)\mathrm{d}m, \tag{9.57}$$

and

$$\int_{j^\circ}^{j} dj = \ln(m/m^\circ) - \int_{m^\circ}^{m} \frac{j}{m} dm - \frac{1}{iK_t} \int_{m^\circ}^{m} \frac{1}{m} \left(\frac{d\Delta T_m}{dm} \right) dm. \quad (9.58)$$

It can be seen that on combining 9.56 and 9.58 the terms which become infinite as $m^\circ \to 0$ will cancel. There is now no difficulty in taking the lower limit of integration to be $m^\circ = 0$, and we note that, as the solution is ideal in this limit $(\Delta T_m/iK_t m) = 1$, hence $j^\circ = 0$. We therefore arrive at the expression

$$\ln \gamma_\pm = -j - \int_0^m \frac{j}{m} dm, \quad (9.59)$$

which poses no problems for a graphical solution. Moreover, G. N. Lewis showed empirically that j was related to m by the expression

$$j = \beta m^\alpha, \quad (9.60)$$

where α and β are constant parameters of the system so that, once α and β have been determined, the integral in 9.55 can be evaluated to give

$$\ln \gamma_\pm = -(\beta + \beta/\alpha)m^\alpha. \quad (9.61)$$

The most direct method of measuring the activity coefficients of electrolytes is by e.m.f. measurements, but as this topic is to be covered in the next chapter we will give no details here. It should be emphasized that where tests have been made there is excellent agreement between activity coefficients obtained from colligative and e.m.f. measurements, and this is surely one of the most satisfying results of applied thermodynamics. Figure 9.10 shows some early data collected by Lewis and Randall for the activity coefficient of sulphuric acid at 298 K.

Table 9.7 gives the values of some activity coefficients in water at 298 K. The Debye–Hückel limiting-law values, shown in parentheses, are reasonably accurate for 1:1 electrolytes up to concentrations of $0.01 \, \text{mol kg}^{-1}$. For higher-valence electrolytes, however, there are significant errors even at $0.001 \, \text{mol kg}^{-1}$.

One of the most satisfactory aspects of the Debye–Hückel theory is that it treats the solvent as a continuous dielectric, but requires no adjustable parameters to get agreement with experiment in the dilute solution limit. Confirmation of this part of the model can be obtained from measurements of the activity coefficient in non-aqueous solvents. Figure 9.11 shows data for HCl in methanol; there is excellent agreement with the limiting law below $0.1 \, \text{mol kg}^{-1}$.

The first correction to the Debye–Hückel limiting law that is usually made is to allow for the finite size of the ions. It is shown in Appendix 9 (A9.11 and A9.16) that if a is the mean radius of the ions then there is an additional factor $r_D \exp(a/r_D)/(a + r_D)$ in the expression for the potential due to the ion and its atmosphere. This factor multiplies only the exponential term in

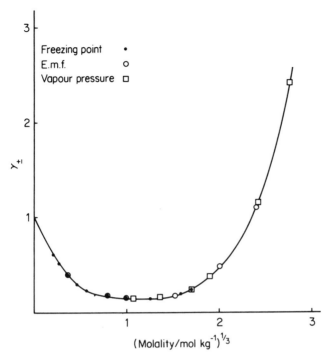

Figure 9.10. The mean ionic activity coefficient of H_2SO_4 in water at 298 K as determined by freezing point, vapour pressure and e.m.f. measurements. (G. N. Lewis and M. Randall, *Thermodynamics*, McGraw-Hill, Maidenhead (1923))

Table 9.7. Mean ionic activity coefficients $\gamma_\pm$ in water at 298 K

$m/\text{mol kg}^{-1}$	HCl	NaCl	H_2SO_4	Na_2SO_4
0.001	0.966	0.964	0.830	
	(0.963)	(0.963)	(0.880)	
0.005	0.929	0.927	0.639	
0.01	0.905	0.902	0.544	
	(0.889)	(0.889)	(0.666)	
0.05	0.830	0.820	0.340	
0.1	0.796	0.778	0.266	0.452
	(0.690)	(0.690)	(0.277)	(0.277)
0.5	0.757	0.681	0.156	0.270
1.0	0.809	0.657	0.132	0.204
2.0	1.009	0.668	0.128	0.154
4.0	1.762	0.783	0.170	0.138

Data from R. A. Robinson and R. H. Stokes, *Electrolyte Solutions*, 2nd Edn revised, Butterworths, London (1970). The values in parentheses are from the Debye–Hückel limiting law, equation 9.50.

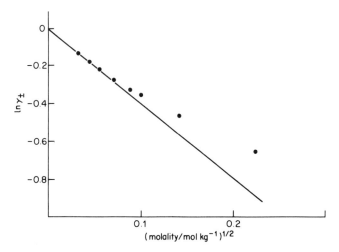

Figure 9.11. The mean ionic activity coefficient for HCl in methanol (298 K). The straight line shows the Debye–Hückel limiting law. (Data from B. E. Conway, *Electrochemical Data*, Elsevier, London and Amsterdam (1952))

the expression for the potential of the ionic atmosphere (9.39) and evaluating this potential at $r = a$ we have

$$\phi'(a) = \frac{-Ze}{4\pi\varepsilon_0\varepsilon_r}\left(\frac{1}{a + r_D}\right). \tag{9.62}$$

Moreover, as the potential anywhere inside a spherical shell of charge is a constant, this expression for ϕ' applies for $0 \leqslant r \leqslant a$.

If we now carry through the charging procedure as in 9.41 the expressions 9.42 and 9.48 are changed only by replacing r_D in the denominator by $(a + r_D)$. For $r_D \gg a$ this is equivalent to multiplying 9.49 by the factor $(1 - a/r_D)$. As r_D is proportional to $I^{-1/2}$ this gives a correction factor to the Debye–Hückel limiting law proportional to $I^{1/2}$ and to a. Although values of a have been determined by comparing this theory with experimental deviations from the limiting law, there are many other factors which contribute to such deviations, and it is probably best to consider the effect of ion size as an illustration that the activity coefficient should obey an expression†

$$\ln \gamma_{\pm} = -|Z_+ Z_-|(PI^{1/2} - QI), \tag{9.63}$$

where P is given by the limiting law and Q is an empirical constant for the system under study. This expression gives a good representation of the initial deviation from the limiting law, perhaps for concentrations of 1:1 electrolytes up to 1 mol kg^{-1}.

---

†H. S. Harned and B. B. Owen, *The Physical Chemistry of Electrolyte Solutions*, Van Nostrand–Reinhold, New York (1958). R. M. Fuoss, *Chem. Rev.*, **17**, 27 (1935).

There is a vast literature dealing with the theory of electrolyte solutions beyond the Debye–Hückel model. Some of these are concerned with mathematics alone, that is they retain the same model as Debye and Hückel and attempt solutions of the equations valid over a greater range of electrolyte concentration. Others extend the range by allowing for ion pairing as well as the Debye–Hückel generalized ion interaction; the best known of such approaches was made by Bjerrum and Fuoss.†

The statistical–mechanical theories of liquids by virial expansions and integral equations (see Chapter 3) have also been applied to electrolyte solutions. The long-range nature of the Coulomb potential is a particular difficulty in this work and also in computer simulations.

Monte Carlo computer simulations have been carried out with the Debye–Hückel assumptions that the ions are hard spheres and the solvent is a continuous dielectric.‡ Figure 9.12 shows the activity coefficients obtained from these computer experiments for a 1:1 electrolyte in water at 298 K and an ionic radius of 4.25 Å. The results extend to concentrations far beyond the Debye–Hückel limit, and a comparison with Figure 9.10 shows that they are in general agreement with experimental observations. In addition they confirm that the solutions obtained from integral equations with the same model are generally satisfactory, particularly from the so-called hyper-netted-chain equations mentioned briefly in Chapter 3.4 of Card and Valleau.‡

From the Monte Carlo calculations it is possible to determine radial distribution functions for both like and unlike ion correlation. Figure 9.13

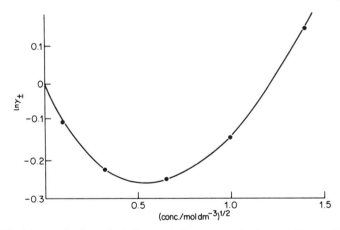

Figure 9.12. Monte Carlo calculations of the mean ionic activity coefficient. A simulation of hard-sphere ions in water at 298 K. D. N. Card and J. P. Valleau‡.

†See, for example, E. A. Guggenheim and R. H. Stokes, *Equilibrium Properties of Aqueous Solutions of Single Strong Electrolytes*, Pergamon Press, Oxford (1969).
‡D. N. Card and J. P. Valleau, *J. Chem. Phys.*, **52**, 6232 (1970); J. C. Rasaiah, D. N. Card and J. P. Valleau, *J. Chem. Phys.*, **56**, 248 (1972).

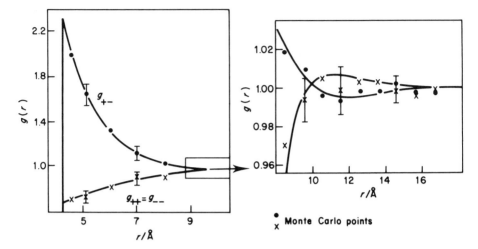

Figure 9.13. Pair correlation functions for like and unlike ions calculated by a Monte Carlo computer simulation. Error bars are for ±1 standard deviation. The lines show the results obtained by solving an integral equation. D. N. Card and J. P. Valleau, see reference on p. 217.

shows the results for a fairly concentrated solution of $1.968 \text{ mol dm}^{-3}$ with agreement again with the integral-equation results. A point of particular interest is that in the region beyond 10 Å there is a crossing of the g_{++} and g_{+-} curves, showing that there is a long-range charge oscillation in the ionic atmosphere. This is expected for concentrated solutions, as explained in the discussion centred around Figure 9.9, but is not predicted by Debye–Hückel theory.

Oscillations in g_{++} have been determined from neutron diffraction studies of concentrated $NiCl_2$ solutions,† but these arise from the hydration-shell structure which has been mentioned earlier. Oscillations of this kind could not arise from any theory of electrolytes which assumes that the solvent is a continuous dielectric. There have been a few attempts to simulate the structure of electrolyte solutions with an assembly of ions and water molecules, but such simulations are computationally very demanding.‡

APPENDIX 9

In this appendix we derive an expression for the functional form of the potential due to an ion and its ionic atmosphere.

The electrostatic potential $\phi(r)$ due to a continuous charge distribution $\rho(r)$ is given by Poisson's equation

---

† J. E. Enderby and G. W. Neilson, *Water, a Comprehensive Treatise*, Ch. 1, Vol. 6, Ed. F. Franks, Plenum Press, London (1979).

‡ See for example J. Fromm, E. Clementi and R. O. Watts, *J. Chem. Phys.*, **78**, 1509 (1974).

$$\nabla^2 \phi(r) = -\rho(r)/\varepsilon_0\varepsilon_r, \tag{A9.1}$$

where

$$\nabla^2 = \partial^2/\partial x^2 + \partial^2/\partial y^2 + \partial^2/\partial z^2. \tag{A9.2}$$

If $\rho(r)$ is spherically symmetrical, $\phi(r)$ is only a function of the radial distance r and satisfies the equation

$$\frac{1}{r^2}\frac{d}{dr}\left(r^2\frac{d\phi}{dr}\right) = -\rho/\varepsilon_0\varepsilon_r \tag{A9.3}$$

To solve this we need another relationship between ϕ and ρ. Debye and Hückel proposed that the charge density at any point should be governed by the Boltzmann distribution law so that $n_j(r)$, the number of ions of type j per unit volume, is given by

$$n_j(r) = \bar{n}_j \exp(-Z_j e\phi/kT), \tag{A9.4}$$

where $\bar{n}_j$ is the average number of ions of type j per unit volume. The charge density is then equal to the concentration of ions multiplied by the charge each one carries. If the solution contains only two types of ion, we have

$$\rho(r) = n_+(r)Z_+e + n_-(r)Z_-e,$$
$$= \bar{n}_+Z_+e\exp(-Z_+e\phi/kT) + \bar{n}_-Z_-e\exp(-Z_-e\phi/kT) \tag{A9.5}$$

where Z_+ is a positive number, Z_- is negative. The average concentration of ions can be expressed in terms of their molality by

$$n_j = m_j\mathcal{N}d, \tag{A9.6}$$

where d is the density of the solvent and $\mathcal{N}$ is Avogadro's number.

Substituting A9.5 into A9.3 gives the so-called Poisson-Boltzmann (P–B) equation

$$\frac{1}{r^2}\frac{d}{dr}\left(r^2\frac{d\phi}{dr}\right) = -\frac{\mathcal{N}de}{\varepsilon_0\varepsilon_r}(m_+Z_+\exp(-Z_+e\phi/kT) + m_-Z_-\exp(-Z_-e\phi/kT)). \tag{A9.7}$$

For a 1:1 electrolyte, $Z_+ = -Z_- = 1$, $m_+ = m_- = m$, and A9.7 becomes

$$\frac{1}{r^2}\frac{d}{dr}\left(r^2\frac{d\phi}{dr}\right) = \frac{\mathcal{N}dem}{\varepsilon_0\varepsilon_r}\sinh(e\phi/kT). \tag{A9.8}$$

Debye and Hückel solved A9.7 approximately by expanding the exponentials and taking the first non-zero term. Because of electrical neutrality

$$m_+Z_+ + m_-Z_- = 0, \tag{A9.9}$$

hence A9.7 becomes

$$\frac{1}{r^2}\frac{d}{dr}\left(r^2\frac{d\phi}{dr}\right) = \frac{\mathcal{N}de^2\phi}{\varepsilon_0\varepsilon_r kT}(m_+Z_+^2 + m_-Z_-^2) = \left(\frac{2\mathcal{N}^2de^2I}{\varepsilon_0\varepsilon_r RT}\right)\phi, \tag{A9.10}$$

where I is the ionic strength (9.21). A9.10 is referred to as the linearized P–B equation; it is a second-order differential equation whose general solution is

$$\phi = \frac{A}{r}e^{-r/r_D} + \frac{B}{r}e^{r/r_D}. \tag{A9.11}$$

This can be confirmed by differentiating A9.11 and substituting the result back into A9.10. When this is done one can equate coefficients of ϕ on both sides of the equation to give the result

$$r_D = \left[\frac{\varepsilon_0 \varepsilon_r RT}{2d\mathcal{N}^2 e^2 I} \right]^{1/2}, \tag{A9.12}$$

which was previously quoted at 9.23.

The constants A and B in A9.11 are determined by boundary conditions on ϕ. First, ϕ must become zero at $r = \infty$ hence $B = 0$. Second, the total charge in the ionic atmosphere must be equal but opposite in sign to the charge on the central ion. If the ions are taken to be rigid spheres of mean radius a (only the close approach of ions of opposite charge is important, hence a can be taken as half their distance of closest approach) then, on integrating the spherical shell of charge $4\pi r^2 \rho$ from a to infinity, we have

$$\int_a^\infty 4\pi r^2 \rho(r)\, dr = -Ze. \tag{A9.13}$$

Substituting for $\rho(r)$ from A9.3 gives

$$4\pi \varepsilon_0 \varepsilon_r \int_a^\infty \frac{d}{dr}\left(r^2 \frac{d\phi}{dr} \right) dr = Ze, \tag{A9.14}$$

which on integration leads to

$$\left| r^2 \frac{d\phi}{dr} \right|_a^\infty = \frac{Ze}{4\pi \varepsilon_0 \varepsilon_r}. \tag{A9.15}$$

Introducing the expression for $d\phi/dr$ from A9.11 ($B = 0$) leads to the result

$$A = \frac{Ze}{4\pi \varepsilon_0 \varepsilon_r}\left(\frac{r_D}{a + r_D} \right) \exp\left(a/r_D \right). \tag{A9.16}$$

If a is small compared with r_D the term in brackets can be replaced by unity to give the result

$$\phi = \frac{Ze}{4\pi \varepsilon_0 \varepsilon_r} \frac{e^{-r/r_D}}{r}, \tag{A9.17}$$

which was quoted in 9.18 and 9.20.

There has been a great deal of discussion of the validity of the linear approximation to the P–B equation A9.10. For the exponents in A9.7 to be less than unity implies that the electrostatic energy $Ze\phi$ is less than the thermal energy, kT. This will be true at large distances from the central ion but the linearized equation is to be solved over the whole range of r from a to infinity and at the short end of the range the assumption will not be valid.

A numerical solution of A9.7 or A9.8 is quite straightforward using the same boundary conditions as for the linearized solution. This was first obtained by Müller in 1927 and extended later by Guggenheim.[†] However, it has also been argued that the faults lie earlier in the P–B equation itself. Onsager was the first to point out that the chance of finding ions i and j a distance r apart would be given by the Boltzmann distribution law either as proportional to $\exp\left(-Z_i e\phi_j/kT \right)$ or as proportional to $\exp\left(-Z_j e\phi_i/kT \right)$. This implies that

$$\frac{\phi_i(r)}{Z_i} = \frac{\phi_j(r)}{Z_j}, \tag{A9.18}$$

a condition which is satisfied by the Debye–Hückel approximate solution, but not by exact solutions of the P–B equation.

---

†E. A. Guggenheim, *Disc. Faraday Soc.*, **24**, 53 (1957).

Chapter 10

Chemical Equilibria in Solution

10.1 PRINCIPLES OF CHEMICAL EQUILIBRIA

In Sections 4.1 and 6.1 we discussed the criteria for the equilibrium between phases. In this chapter we deal with chemical equilibria in solution, with particular emphasis on the solubility of electrolytes and the strengths of acids and bases. The basic criterion for chemical equilibrium is the same as that for phase equilibrium, namely, the Gibbs free energy of the system shall be a minimum. For phase equilibria this is equivalent to the important result that the chemical potential of a chemical species must be the same in each phase. For chemical equilibria there is a different criterion which relates the concentrations of each chemical species to differences in their chemical potentials in their standard states.

We consider a general stoichiometric chemical reaction

$$aA + bB + \ldots \rightarrow mM + nN + \ldots. \qquad (10.1)$$

If it proceeds from left to right by a small amount, insufficient to change concentrations of reactants and products, then the free energy change per mole of reaction is

$$\Delta G_m = m\mu_m + n\mu_n \ldots - a\mu_a - b\mu_b \ldots = \sum_p k_p\mu_p - \sum_r k_r\mu_r, \qquad (10.2)$$

where k_p and k_r are the stoichiometric numbers of product and reactant molecules respectively.

The chemical potential is a function of temperature, pressure and the concentrations of other species in the phase. We have earlier met expressions for the chemical potential of a perfect gas (4.25), an imperfect gas (4.26) for the components of an ideal liquid mixture (6.39), for the solvent and solute in non-ideal liquids (6.67) and (6.75), and for electrolytes (9.25). These expressions all have the same form which we shall generalize to

$$\mu = \mu^\ominus + RT \ln a, \qquad (10.3)$$

where $\mu^{\ominus}$ is the chemical potential of a standard state and a is the activity.

It should be recalled that to complete the definition of activity it is necessary to specify the relationship between a and some measure of the composition of the system. In an ideal system a will be identical to pressure measured in atmospheres, or mole-fraction, or molality measured in moles per kg of solvent, and in non-ideal systems it will approach these numbers as ideality is approached (e.g. at zero pressure or infinite dilution).

On substituting 10.3 into 10.2 we have

$$\Delta G_m = \left(\sum_p k_p \mu_p^{\ominus} - \sum_r k_r \mu_r^{\ominus}\right) + RT\left(\sum_p k_p \ln a_p - \sum_r k_r \ln a_r\right), \quad (10.4)$$

which can be simplified to

$$\Delta G_m = \Delta G_m^{\ominus} + RT \ln\left[\frac{\prod a_p^{k_p}}{\prod a_r^{k_r}}\right], \quad (10.5)$$

where $\prod$ represents the product of the activities raised to powers of their stoichiometric numbers. $\Delta G_m^{\ominus}$ is the free energy change per mole of reaction† with all species in their standard states.

$$\Delta G_m^{\ominus} = \sum_p k_p \mu_p^{\ominus} - \sum_r k_r \mu_r^{\ominus} \quad (10.6)$$

Expression 10.5 is known as the van't Hoff isotherm, or general reaction isotherm.

If the reaction is at equilibrium then for any further infinitesimal change ΔG_m will be zero. In this case we have from 10.5

$$\Delta G_m^{\ominus} = -RT \ln\left[\frac{\prod a_p^{k_p}}{\prod a_r^{k_r}}\right]. \quad (10.7)$$

The ratio of products of activities at equilibrium which appears on the right-hand side of this equation is called the thermodynamic equilibrium constant. We write

$$K_a = \frac{\prod a_p^{k_p}}{\prod a_r^{k_r}}, \quad (10.8)$$

where

$$\Delta G_m^{\ominus} = -RT \ln K_a. \quad (10.9)$$

K_a is a constant at constant T because $\Delta G_m^{\ominus}$ is an energy defined with respect to all species in their standard states. $\Delta G_m^{\ominus}$ for the reaction can be calculated by reference to standard thermodynamic tables which record free energies of formation of the chemical species involved (generally at 298 K).

---

†If any stoichiometric number differs from unity, it is necessary to specify the particular reactant or product serving as reference.

Chemists are interested in knowing the ratios of concentrations of species present at equilibrium so that expressions 10.8 and 10.9 are not very useful for non-ideal systems unless we have a method of relating activities to concentrations. For this reason one chooses to define the composition of a chemical system in units such that the activity coefficients shall be most nearly equal to unity. For example, for gases the most suitable unit is the numerical value of the pressure in atmospheres because for non-ideal gases pressure is almost equal to fugacity and the activity is fugacity measured in atmospheres. As a result K_p defined by

$$K_p = \frac{\prod (P_p/\text{atm})^{k_p}}{\prod (P_r/\text{atm})^{k_r}}, \tag{10.10}$$

will be a number which is almost independent of total pressure for most gases.

For non-electrolyte liquid mixtures, the mole fraction is the best concentration unit because activity is approximately equal to mole fraction and the ratio

$$K_x = \frac{\prod x_p^{k_p}}{\prod x_r^{k_r}}, \tag{10.11}$$

is approximately independent of composition. For dilute solutions of electrolytes, molality is the most widely adopted unit and

$$K_m = \frac{\prod (m_p/\text{mol kg}^{-1})^{k_p}}{\prod (m_r/\text{mol kg}^{-1})^{k_r}}, \tag{10.12}$$

is approximately constant. However, deviations from ideality are much greater for electrolyte solutions than for non-electrolytes, as we have explained in Chapter 9.

We will illustrate the use of some of these equilibrium constants by considering the equilibrium

$$N_2O_4 \rightleftharpoons 2NO_2, \tag{10.13}$$

in both the gas phase and in solution. From standard thermodynamic tables we find the following free energies of formation at 298 K:

$$NO_2(g) \quad \Delta G_f^{\ominus} = -38.4 \text{ kJ mol}^{-1}$$
$$N_2O_4(g) \quad \Delta G_f^{\ominus} = -81.6 \text{ kJ mol}^{-1}$$
$$N_2O_4(l) \quad \Delta G_f^{\ominus} = -81.9 \text{ kJ mol}^{-1}$$

Thus in the gas phase at 298 K

$$K_p = e^{-\Delta G_m^{\ominus}/RT} = e^{+(2\times 38.4 - 81.6)/2.48} = 0.14. \tag{10.14}$$

In the pure liquid there is very little NO_2 and no accurate measurements of the equilibrium constant have been made. However, in CCl_4 at 298 K experiment has shown that the solution is quite close to ideal and a value of the equilibrium constant has been obtained

$$K_x = 6.4 \times 10^{-5}. \tag{10.15}$$

The two equilibrium constants 10.14 and 10.15 have very different values and it would be interesting to use them to compare the ratios of NO_2 and N_2O_4 in the two phases. This cannot be done directly because this ratio depends on the mole fractions or total pressures. However, if gas and solution are in equilibrium we can write

$$\mu_{N_2O_4}(sol) = \mu_{N_2O_4}(g). \tag{10.16}$$

Assuming ideality in both phases we have

$$\mu^{\ominus}_{N_2O_4}(l) + RT \ln x = \mu^{\ominus}_{N_2O_4}(P = 1) + RT \ln (P_{N_2O_4}/atm). \tag{10.17}$$

If the mole fraction of N_2O_4 in CCl_4 solution is 0.1 then the relevant numbers in 10.17 are

$$-81.9 + 2.48 \ln 0.1 = -81.6 + 2.48 \ln (P_{N_2O_4}/atm), \tag{10.18}$$

whence

$$P_{N_2O_4} = 0.089 \text{ atm.} \tag{10.19}$$

In this case we have in solution

$$\frac{x^2_{NO_2}}{0.1} = 6.4 \times 10^{-5}, \quad \therefore x_{NO_2} = 2.5 \times 10^{-3}, \tag{10.20}$$

and

$$\frac{x_{NO_2}}{x_{N_2O_4}} \approx \frac{[NO_2]}{[N_2O_4]} = 2.5 \times 10^{-2}. \tag{10.21}$$

In the gas phase

$$\frac{P^2_{NO_2}}{0.089} = 0.14, \quad \therefore P_{NO_2} = 0.11 \text{ atm,} \tag{10.22}$$

and

$$\frac{P_{NO_2}}{P_{N_2O_4}} = \frac{[NO_2]}{[N_2O_4]} = 1.24. \tag{10.23}$$

It can be seen from these figures that the ratio of NO_2 to N_2O_4 is much greater in the gas phase than in solution, a result readily explained by the fact that the decomposition of N_2O_4 in the gas phase is favoured by the extra translational entropy of the NO_2.

In the gas phase there is usually no difficulty in defining what are the chemical species. In the above example we have considered an equilibrium between NO_2 and N_2O_4; at higher temperatures we might have to include NO and O_2 arising from the reaction

$$2NO_2 \rightleftharpoons 2NO + O_2. \tag{10.24}$$

All of these species can be identified unambiguously and their concentrations determined by spectroscopic or other means.

In the liquid phase the identification of chemical species is less easy. For example, there might be spectroscopic evidence to show that the NO_2 is strongly solvated. If this is so, should we distinguish between free NO_2 in solution and solvated NO_2; is there more than one type of solvation complex? In aqueous solutions the situation is often more complex because of the strong hydrogen bonding capability of water and its cooperative effect. Thus a species A could be hydrated to a multiplicity of forms $A(H_2O)_n$ which would generally be exchanging H_2O rapidly with the solvent.

It is unlikely that all feasible species that could be present in solution can be identified by physical or chemical means. If this is so there is no chance of determining the multiplicity of equilibrium constants that would fully describe the equilibrium system. Accepting this, the empirical approach is to treat the system as a simple equilibrium between NO_2 and N_2O_4 at the expense of having a large deviation from ideality. In short, large deviations from ideality are a reflection of the fact that other equilibria arising from solvation, hydrogen bonding, etc. are occuring in the solution.

10.2   EQUILIBRIUM ELECTROCHEMISTRY

The field of electrochemistry, particularly non-equilibrium electrochemistry, is too wide to be contained within the format of this book. Nevertheless a brief introduction to equilibrium electrochemistry is required for its relevance to the determination of the activities of ions in polar, mainly aqueous, solvents.

The chemical reaction represented by

$$Zn + Cu^{2+}(aq) \rightarrow Zn^{2+}(aq) + Cu, \tag{10.25}$$

can be carried out in two contrasting ways. The first is to add powdered zinc to copper sulphate solution. The zinc will dissolve and copper will be precipitated. The reaction is being carried out irreversibly and heat will be evolved equal to $-\Delta H = 217 \text{ kJ mol}^{-1}$ per mole of reactant.

A second method is to combine the reagents so as to construct an electrochemical cell, called the Daniell cell, one electrode of which is a zinc rod dipped in an aqueous solution of $ZnSO_4$ and the other is a copper rod dipped in an aqueous solution of $CuSO_4$. The two solutions are connected by an electrically conducting bridge which prevents the diffusion of Cu^{2+} and Zn^{2+} between the two cell compartments: this is typically a concentrated solution of K_2SO_4 in an agar gel which is known as a salt bridge. The cell is represented in a simplified form in Figure 10.1.

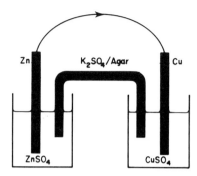

Figure 10.1. Simple representation of the Daniell cell. The direction of electron flow under short circuit is shown

If the two electrodes are short circuited then the reaction represented by 10.25 will occur, the zinc electrode will dissolve, and copper will be deposited on the copper electrode. Electrons will flow between the electrodes in the direction indicated and negative ions will flow through the salt bridge to maintain electrical neutrality.

The cell can also be run reversibly by applying an external e.m.f. $\mathscr{E}$, across the two electrodes which just balances the natural e.m.f. of the cell reaction. By changing this external e.m.f. slightly, either increasing or decreasing $\mathscr{E}$, current will dribble through the cell in one direction or the other. The chemical reaction can in this way be made to proceed forwards or backwards in a thermodynamically reversible manner.

The electric current extracted from a cell can be directly converted to work through an electric motor and, as the system is at constant pressure and temperature, the work available when the cell is operating reversibly measures the Gibbs free energy of the reaction. One mole of reaction produces nF Coulombs of charge through the circuit, n being the number of moles of electrons per mole of reaction ($n = 2$ for reaction 10.25) and F is the Faraday, $F = 96,487$ Coulombs. This charge passes against an external e.m.f. $\mathscr{E}$ and produces an amount of work $nF\mathscr{E}$. As ΔG_m is the work done on the system in a reversible process at constant T and P, we have

$$\Delta G_m = -nF\mathscr{E}. \tag{10.26}$$

The sign of ΔG_m for reaction 10.25 is given by convention as

$$\Delta G_m = G_{\text{products}} - G_{\text{reactants}}, \tag{10.27}$$

and by virtue of the relationship 10.26 we require a consistent convention for the sign of the cell e.m.f. The cell which carries out the reaction 10.25 is written symbolically as

$$Zn|Zn^{2+}\|Cu^{2+}|Cu, \tag{10.28}$$

which we can read from left to right as Zn losing electrons to given Zn^{2+} and

Cu^{2+} gaining electrons to give Cu. Exactly the same words can be applied to equation 10.25 as written. The double bar in the middle of 10.28 indicates that there is no contribution to the e.m.f. arising from contact between the two solutions: the salt bridge eliminates what is called the liquid junction potential which is a source of irreversibility, and the single vertical lines indicate phase boundaries at the electrodes.

The e.m.f. of the cell 10.28 is determined by the balance between the tendency of Zn to pass into solution as Zn^{2+} and Cu to pass into solution as Cu^{2+}. These tendencies are measured by the e.m.f.s of the so-called half-cells, labelled according to their positions on the left (suffix L) or the right (suffix R) of the cell as written in 10.28,

$$Zn^{2+}|Zn \qquad \mathscr{E}_L \qquad (10.29)$$

$$Cu^{2+}|Cu \qquad \mathscr{E}_R \qquad (10.30)$$

for which the appropriate chemical reactions and free energies are

$$Zn^{2+} + 2e \rightarrow Zn \quad \Delta G_L \qquad (10.31)$$

$$Cu^{2+} + 2e \rightarrow Cu \quad \Delta G_R \qquad (10.32)$$

The chemical reaction 10.25 can be constructed by taking $10.32 - 10.31$, hence ΔG_m for this reaction is given by

$$\Delta G_m = \Delta G_R - \Delta G_L. \qquad (10.33)$$

Likewise the cell 10.28 can be obtained by combining the two half cells $10.30 - 10.29$ so that the e.m.f. of the cell is related to the half cell e.m.f.s by

$$\mathscr{E} = \mathscr{E}_R - \mathscr{E}_L. \qquad (10.34)$$

We emphasize that the sign of the e.m.f. of a cell refers to the symbolic representation such as 10.28 rather than to the actual cell as set up on the laboratory bench. Thus the e.m.f. of

$$Cu|Cu^{2+}\|Zn^{2+}|Zn, \qquad (10.35)$$

has the opposite sign to 10.28 although it is the same laboratory cell. However, if the standard electrode potentials (see below) are known, it is customary to arrange the cell so that its standard e.m.f. is positive.

Standard Reduction Potentials

Expression 10.5 embodies the dependence of the Gibbs free energy on the activities of reactants and products. Because of the proportionality between $\mathscr{E}$ and ΔG_m, the e.m.f. of the cell must also depend on the activities of the cell components. By substituting 10.26 into 10.5 we obtain the Nernst equation

$$\mathscr{E} = \mathscr{E}^{\ominus} - \frac{RT}{nF} \ln \left[\frac{\prod a_p^{k_p}}{\prod a_r^{k_r}} \right] \qquad (10.36)$$

where $\mathscr{E}^{\ominus}$ is the e.m.f. of the cell, all components at unit activity. In the example under consideration, this is the cell

$$Zn|Zn^{2+}(a = 1)\|Cu^{2+}(a = 1)|Cu, \qquad (10.37)$$

the free energy of the cell reaction being

$$\Delta G_m^{\ominus} = -2F\mathscr{E}^{\ominus}. \qquad (10.38)$$

Note that the activities of solids, such as Zn and Cu in the present case, are always taken to be unity, the implication being that their chemical potentials are constant.

We have seen from 10.34 that the e.m.f. of a cell is equal to the difference in the e.m.f.s of two half-cells. If the half-cells contain ions of unit activity then this difference gives $\mathscr{E}^{\ominus}$

$$\mathscr{E}^{\ominus} = \mathscr{E}_R^{\ominus} - \mathscr{E}_L^{\ominus}. \qquad (10.39)$$

A table of standard half-cell potentials can therefore be constructed provided a suitable zero is chosen; absolute values of half-cell potentials have no meaning because two half-cells are required to obtain a practical cell with a measurable e.m.f.

The reference zero half-cell potential is chosen to be consistent with the zero Gibbs free energy of ions in solution which, as described in Chapter 9, is $\Delta G_m(H^+(a = 1)) = 0$. Thus the standard half-cell with zero e.m.f. is

$$Pt|H_2(1 \text{ atm})|H^+(a = 1), \ \mathscr{E}^{\ominus} = 0. \qquad (10.40)$$

To be precise, the hydrogen gas should also be at unit activity which is roughly one atmosphere pressure. This standard electrode can be constructed by bubbling H_2 at 1 atm around a platinized platinum electrode dipped in acid with unit hydrogen ion activity. The platinum acts as a catalyst for the ionization of the hydrogen, which is otherwise a slow process.

Table 10.1 lists a few standard electrode potentials. In principle, these can all be taken directly as the e.m.f. of a cell whose left-hand electrode is the standard hydrogen electrode. For example,

$$Pt|H_2(1 \text{ atm})|HCl \ (a = 1)|AgCl(s)|Ag \qquad (10.41)$$

is a cell with a silver–silver chloride electrode (Ag with a coating of AgCl dipped in HCl) on the right. The e.m.f. of this cell is, from Table 10.1,

$$\mathscr{E}^{\ominus} = 0.2223 \text{ V}. \qquad (10.42)$$

The reaction associated with this cell is

$$H_2 + AgCl \rightleftharpoons 2H^+ + 2Cl^- + Ag,$$

and as this is a two-electron process, i.e. a process involving two moles

Table 10.1. Standard reduction potentials in acid solution at 298 K

Cell reaction	$\mathscr{E}^{\ominus}/V$
$Li^+ + e \rightleftharpoons Li$	-3.041
$Na^+ + e \rightleftharpoons Na$	-2.71
$\frac{1}{2}Mg^{2+} + e \rightleftharpoons \frac{1}{2}Mg$	-2.372
$\frac{1}{3}Al^{3+} + e \rightleftharpoons \frac{1}{3}Al$	-1.662
$\frac{1}{2}V^{2+} + e \rightleftharpoons \frac{1}{2}V$	-1.175
$\frac{1}{2}Zn^{2+} + e \rightleftharpoons \frac{1}{2}Zn$	-0.763
$\frac{1}{2}Fe^{2+} + e \rightleftharpoons \frac{1}{2}Fe$	-0.447
$D^+ + e \rightleftharpoons \frac{1}{2}D_2(g)$	-0.0034
$H^+ + e \rightleftharpoons \frac{1}{2}H_2(g)$	0
$Cu^{2+} + e \rightleftharpoons Cu^+$	0.153
$AgCl + e \rightleftharpoons Ag + Cl^-$	0.2223
$\frac{1}{2}Cu^{2+} + e \rightleftharpoons \frac{1}{2}Cu$	0.3419
$Fe^{3+} + e \rightleftharpoons Fe^{2+}$	0.771
$Ag^+ + e \rightleftharpoons Ag$	0.7996
$\frac{1}{2}Br_2(l) + e \rightleftharpoons Br^-$	1.066
$H^+ + \frac{1}{4}O_2(g) + e \rightleftharpoons \frac{1}{2}H_2O(l)$	1.229
$Ce^{4+} + e \rightleftharpoons Ce^{3+}$	1.61
$\frac{1}{2}S_2O_8^{2-} + e \rightleftharpoons SO_4^{2-}$	2.010

Data from CRC *Handbook of Chemistry and Physics*, 69th Edition, 1988.

of electrons per mole of AgCl reacted, the standard Gibbs free energy change is

$$\Delta G_m^{\ominus} = -2F\mathscr{E}^{\ominus}. \tag{10.43}$$

Table 10.1 has been written with all reactions as one-electron reductions. This has the advantage that there is linear relationship between $\mathscr{E}^{\ominus}$ and $\Delta G_m^{\ominus}$. If the reader keeps in mind the fact that $-F\mathscr{E}^{\ominus}$ is a free-energy scale the difficulties frequently met in keeping track of signs should be alleviated.

One of the most useful features of the table of standard reduction potentials is the opportunity it affords to decide which way a redox reaction will go. For example, will Cu reduce Fe^{3+} to Fe^{2+} or will Fe^{2+} reduce Cu^{2+} to Cu? The half-cell reactions in Table 10.1 have been written with oxidizing agents on the left increasing in strength down the page, and reducing agents on the right increasing in strength up the page. The strongest reducing agent has the most negative $\mathscr{E}^{\ominus}$ and the strongest oxidizing agent the most positive $\mathscr{E}^{\ominus}$.

If we compare any pair of reactions in the table then the one above (i.e. that with the less positive $\mathscr{E}^{\ominus}$) will move to the left and the one below (i.e. that with the more positive $\mathscr{E}^{\ominus}$) to the right. Thus Cu will reduce Fe^{3+} to Fe^{2+}. The standard free energy change for the reaction

$$Fe^{3+} + \tfrac{1}{2}Cu \rightarrow Fe^{2+} + \tfrac{1}{2}Cu^{2+}, \tag{10.44}$$

will be obtained from $\mathscr{E}^{\ominus}$ for the cell†

$$Cu|Cu^{2+}\|Fe^{3+}, Fe^{2+}|Pt \tag{10.45}$$

$$\mathscr{E}^{\ominus} = \mathscr{E}_R^{\ominus} - \mathscr{E}_L^{\ominus} = 0.771 - 0.342 = 0.429 \text{ V}$$

$$\Delta G_m^{\ominus} = -F\mathscr{E}^{\ominus} = -41.5 \text{ kJ mol}^{-1} \tag{10.46}$$

Standard reduction potentials refer to half-cells with all species in their standard states and, as has been emphasized in Chapter 9, for ionic species these are hypothetical states extrapolated from ideal (dilute) solutions. Thus $\mathscr{E}^{\ominus}$ cannot be obtained directly by measuring a cell e.m.f. but must be obtained by extrapolation. We will illustrate this from some data on the cell

$$Pt|H_2(1 \text{ atm})|HCl(m)|AgCl|Ag, \tag{10.47}$$

which was studied by Harned and Ehlers in great detail. Their measured e.m.f.s at 298 K are given in Table 10.2.

The cell reaction is

$$\tfrac{1}{2}H_2(1 \text{ atm}) + AgCl \rightarrow Ag + H^+ + Cl^-, \tag{10.48}$$

and from the Nernst equation 10.36 we can write the e.m.f.

$$\mathscr{E} = \mathscr{E}^{\ominus} - \frac{RT}{F} \ln(a_{H^+}a_{Cl^-}), \tag{10.49}$$

the activities of all other species being unity. The mean ionic activity coefficient is defined by

$$a_{H^+}a_{Cl^-} = m^2\gamma_{\pm}^2, \tag{10.50}$$

Table 10.2. Measured e.m.f. of the cell 10.47 at 298 K as a function of HCl molality

m(HCl)	$\mathscr{E}$/V	m(HCl)	$\mathscr{E}$/V
0.005	0.4984	0.04	0.3967
0.006	0.4894	0.05	0.3859
0.007	0.4818	0.06	0.3771
0.008	0.4752	0.07	0.3697
0.009	0.4694	0.08	0.3632
0.01	0.4642	0.09	0.3575
0.02	0.4302	0.1	0.3524
0.03	0.4106		

Data from H. S. Harned and R. W. Ehlers, *J. Am. Chem. Soc.*, **54**, 1350 (1932); **55**, 2179 (1933).

---

†The type of electrode on the right is called an oxidation–reduction electrode and consists of an inert metal collector, usually bright platinum, immersed in a solution which contains ions of the same element in two different oxidation states.

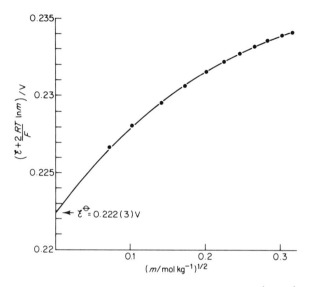

Figure 10.2. Graphical extrapolation to find $\mathscr{E}^\ominus$ for the $Cl^-/AgCl/Ag$ half cell

and substituting this into 10.49 gives

$$\mathscr{E} = \mathscr{E}^\ominus - \frac{2RT}{F}\ln m - \frac{2RT}{F}\ln \gamma_\pm. \tag{10.51}$$

As we approach infinite dilution $\gamma_\pm$ approaches zero but in this limit $\ln m$ approaches minus infinity, so it is difficult to obtain an accurate measure of $\mathscr{E}^\ominus$ without taking into account the dependence of $\gamma_\pm$ on m. This can be done using the Debye–Hückel limiting law. For a 1:1 electrolyte in water at 298 K we have, from 9.50

$$\ln \gamma_\pm = -1.172m^{1/2}. \tag{10.52}$$

A graph of $\mathscr{E} + 2(RT/F)\ln m$ against $m^{1/2}$ is shown in Figure 10.2. The extrapolated value of $\mathscr{E}^\ominus$ is equal to the standard reduction potential for the AgCl/Ag, Cl^- half-cell because the left half-cell in 10.47 is the hydrogen electrode whose $\mathscr{E}^\ominus$ is zero. Once $\mathscr{E}^\ominus$ is known a measurement of $\mathscr{E}$ gives the activity coefficient for HCl; such values were discussed in Chapter 9.

10.3 SOLUBILITY OF SALTS

In dilute aqueous solutions most salts are completely dissociated to ions, and this is true whether the salts are very soluble like NaCl or very insoluble like AgCl. The solubility of a sparingly soluble salt M_pX_q can therefore be analysed through the equilibrium

$$M_pX_q(s) \rightleftharpoons pM^{q+}(aq) + qX^{p-}(aq). \tag{10.53}$$

The activity of a solid is always taken as unity, hence the equilibrium constant (10.8) is a function only of the activities of the ions. This equilibrium constant is called the solubility product and is written

$$K_{sp} = (a_{M^+})^p(a_{X^-})^q = \gamma_{\pm}^{p+q}(m_{M^+})^p(m_{X^-})^q \qquad (10.54)$$

An approximate value of the solubility of the salt can be obtained by taking $\gamma_{\pm} = 1$. If S is the solubility in moles per kg, then

$$m_{M^+} = pS, \ m_{X^-} = qS, \qquad (10.55)$$

and from 10.54

$$K_{sp} \approx (pS)^p(qS)^q \qquad (10.56)$$

whence

$$S \approx \left(\frac{K_{sp}}{p^p q^q}\right)^{\frac{1}{p+q}} \qquad (10.57)$$

To obtain a more accurate value of the solubility one must take into account the value of the activity coefficient $\gamma_{\pm}$; it can be determined by experiment or estimated from the Debye–Hückel limiting law. Alternatively, if K_{sp} can be determined by e.m.f. measurements, ..., and S determined by independent means (e.g. spectroscopic measurements), $\gamma_{\pm}$ can be calculated.

The solubility product can be calculated from the standard free energies of formation of the solid and the ions, or derived from measurements of the e.m.f. of an appropriate cell. For example, the cell

$$Ag|Ag^+ \|Cl^-|AgCl|Ag, \qquad (10.58)$$

has at the right-hand electrode the reaction

$$AgCl + e \rightleftharpoons Ag + Cl^-, \qquad (10.59)$$

and at the left-hand electrode

$$Ag^+ + e \rightleftharpoons Ag. \qquad (10.60)$$

The overall cell reaction is

$$AgCl \rightleftharpoons Ag^+ + Cl^-, \qquad (10.61)$$

and the e.m.f. of the cell will be given by the Nernst equation

$$\mathscr{E} = \mathscr{E}^{\ominus} - \frac{RT}{F} \ln[(a_{Ag^+})(a_{Cl^-})] = \mathscr{E}^{\ominus} - \frac{RT}{F} \ln K_{sp}, \qquad (10.62)$$

where, from Table 10.1

$$\mathscr{E}^{\ominus} = \mathscr{E}_R^{\ominus} - \mathscr{E}_L^{\ominus} = 0.2223 - 0.7996 = -0.5773 \text{ V}. \qquad (10.63)$$

At equilibrium, the cell e.m.f. is zero hence from 10.62 we have

$$\ln K_{sp} = \frac{-0.5773 \times 9.648 \times 10^4}{8.314 \times 298} = -22.48 \qquad (10.64)$$

from which

$$K_{sp}(298 \text{ K}) = 1.72 \times 10^{-10}. \tag{10.65}$$

From expression 10.57 ($p = q = 1$), we can now deduce the solubility of AgCl in water at 298 K to be

$$1.31 \times 10^{-5} \text{ mol kg}^{-1}.$$

If the concentration of either Ag^+ or Cl^- is increased by adding a soluble salt with a common ion, such as $AgNO_3$ or NaCl, the concentration of the other ion must decrease in order to maintain the constancy of K_{sp}. For example, if sufficient NaCl is added to give a 0.1 M solution then, if the solubility of AgCl is S, the concentration of Ag^+ will be S and of Cl^- $(S + 0.1)$. Thus from 10.65

$$S(S + 0.1) = 1.72 \times 10^{-10}, \tag{10.66}$$

and, noting that S is negligible compared to 0.1, this has the solution $S = 1.72 \times 10^{-9}$ mol kg^{-1}. In other words, 0.1 M NaCl decreases the solubility of AgCl by a factor 10^{-4}. This is known as the 'salting-out' effect arising from the presence of a common ion.

A 'salting-in' effect can be achieved by adding a salt *without* a common ion as this will reduce the activity coefficient below unity. For example, a 0.1 M solution of $NaNO_3$ has, from 10.52, an activity coefficient of

$$\ln \gamma_{\pm} = -0.1172, \therefore \gamma_{\pm} = 0.89, \tag{10.67}$$

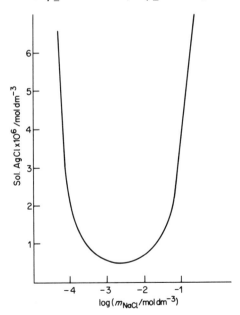

Figure 10.3. Dependence of AgCl solubility on the concentration of added NaCl. (Data for 298 K from F. W. Linke, *Solubilities of Inorganic and Metal Organic Compounds*, D. Van Nostrand, New York (1958))

hence, from 10.54 and 10.65

$$S = K_{sp}^{1/2}/\gamma = 1.47 \times 10^{-5} \tag{10.68}$$

so that the solubility has been increased by 6%.

This is a much smaller change (in the opposite direction) than can be achieved by the common ion effect and, although higher concentrations of added salt can be used, the solution will then be in a concentration range where the Debye–Hückel limiting law breaks down. Indeed at high salt concentrations there are no universal rules. For example, the solubility of AgCl decreases with added chloride ion up to about 10^{-3} M but then undergoes an increase as shown in Figure 10.3. This is not a simple thermodynamic effect, it is probably due to the formation of complex ions $[AgCl_{n+1}]^{n-}$.

10.4 ACID AND BASE STRENGTHS

In 1925 Brönsted defined an acid as a species having a tendency to lose a proton and a base as a species having a tendency to gain a proton; a deprotonated acid is a base and a protonated base is an acid. Thus acids (A) and bases (B) are linked in conjugate acid–base pairs connected schematically by the equilibrium

$$A \rightleftharpoons B + H^+. \tag{10.69}$$

For example, CH_3COOH is an acid and CH_3COO^- a base; NH_4^+ an acid, NH_3 a base. The free proton does not exist in the liquid state and hence the equilibrium 10.69 has only schematic significance. In practice an equilibrium is always established between two conjugate pairs, (A_1, B_1) and (A_2, B_2), as follows

$$A_1 + B_2 \rightleftharpoons B_1 + A_2. \tag{10.70}$$

An example is

$$CH_3COOH + NH_3 \rightleftharpoons CH_3COO^- + NH_4^+. \tag{10.71}$$

If K_1 and K_2 are equilibrium constants for the two schematic equilibria of type 10.69, then the equilibrium constant for 10.70 is

$$K = K_1/K_2. \tag{10.72}$$

Relative scales of acid and base strength can be established for each solvent (these scales will depend slightly on temperature) and it is customary and convenient to use the solvent itself as the standard. For example, in water, acid strength is measured relative to the strength of H_3O^+ by examining the equilibrium

$$A + H_2O \rightleftharpoons B + H_3O^+. \tag{10.73}$$

In dilute solutions the activity of water can be taken as constant and omitted, hence the equilibrium constant for 10.73 is given by

$$K_a = \frac{a_B a_{H_3O^+}}{a_A}, \tag{10.74}$$

and this is called the acid dissociation constant of A. The stronger the acid A, the more it will protonate the water molecule and the greater will be K_a.

A standard of base strength can be established relative to the unprotonated solvent as standard. Thus in water solutions we examine the equilibrium

$$B + H_2O \rightleftharpoons A + OH^-, \tag{10.75}$$

and define base strength by the value of

$$K_b = \frac{a_A \, a_{OH^-}}{a_B}. \tag{10.76}$$

By taking the product of 10.74 and 10.76 we see that acid and base strengths are related:

$$K_a K_b = (a_{H_3O^+})(a_{OH^-}) = K_s, \tag{10.77}$$

where K_s is the equilibrium constant for auto-ionization of the solvent, in this case water. K_w has the value 1.00×10^{-14} at 298 K.

As equilibrium constants can take values covering many powers of ten, it is traditional to use a logarithmic scale to measure acid and base strengths

$$pK = -\log_{10} K. \tag{10.78}$$

This is called the pK scale. Table 10.3 lists some values for acids and bases

Table 10.3. Dissociation constants for some common acids and bases in water at 298 K. For carbonic and phosphoric acids dissociation constants are given for successive ionization of the protons

Acid	pK_a	Base	pK_b
Phenol	9.89	Aniline	9.37
Acetic acid	4.75	Pyridine	8.75
Benzoic acid	4.19	Methylamine	3.34
Formic acid	3.75	Trimethylamine	4.19
Trichloracetic acid	0.70	Ammonium hydroxide	4.75
Hypochlorous acid	7.53	Hydroxylamine	7.97
Hydrofluoric acid	3.45		
Carbonic acid (1)	6.37		
(2)	10.25		
Phosphoric acid (1)	2.12		
(2)	7.21		
(3)	12.67		

Data from CRC *Handbook of Chemistry and Physics*, 59th Edn, 1978–79.

in water at 298 K. Strong acids have $pK_a < -1$ and are more than 90% ionized: nitric acid, for example, has a pK_a of approximately -1.4. Other common strong acids, hydrochloric, sulfuric and perchloric are effectively 100% ionized in water and cannot be differentiated in acid strength. However, we shall see later that a discrimination can be made in solvents which are weaker proton acceptors than water.

pH Measurement

It is important to distinguish between the strength of an acid as measured by the degree to which it protonates the solvent (i.e. pK_a) and its acidity, which is the concentration of hydrogen ions (protonated solvent molecules) in the solution. In 1909 Sörensen proposed that the logarithm of the hydrogen ion concentration be taken as a measure of acidity, and he introduced the term pH defined by

$$pH = -\log[H^+]. \tag{10.79}$$

There are two ways in which this simple definition can be extended. First, we can recognize that protons are attached to solvent molecules and, in the case of water, write

$$pH = -\log[H_3O^+]. \tag{10.80}$$

However, aqueous solutions are not ideal and any thermodynamic measurement will give in the first place activities rather than the concentrations of the ionic species. The most widely accepted definition of pH today is

$$pH = -\log a_{H_3O^+} \tag{10.81}$$

but this encounters the problem that the activities of individual ions are not absolutely determined. It should be recalled that thermodynamic measurements on ionic solutions give the *mean* activity coefficients of the positive and negative ions (9.43).

The procedure recommended by the National Bureau of Standards for the accurate determination of pH according to the recipe 10.81 is to measure the e.m.f. of the cell

$$Pt, H_2(1\ atm)|buffer\ with\ Cl^-|AgCl(s), Ag \tag{10.82}$$

as a function of Cl^-. One can then use certain well established formulae for the activity of Cl^- and determine a_{H^+} from 10.49.

This procedure is rather tedious and the hydrogen electrode is not a very easy standard to use because the catalytic activity of the platinum is subject to poisoning. The most common way to measure pH is with a glass electrode which consists of a Ag|AgCl electrode contained within a thin glass bulb in an HCl solution of known pH. This electrode is immersed in the solution whose pH is required and its e.m.f. measured against some other standard electrode. The glass wall behaves as a membrane which is permeable to hydrogen ions and it has been found that the glass-electrode potential is linearly proportional to the pH of the solution being examined. The glass electrode has also been found useful for measuring the pH in water–organic solvent mixtures providing that they contain more than 10% by weight of water.

For strong acids in aqueous solutions at concentrations appreciably greater than 10^{-7} M, the concentration of H_3O^+ is approximately equal to the concentration of acid. Thus 10^{-3} M HCl has pH ≈ 3. At concentrations less than 10^{-6} M, the amount of H_3O^+ produced by the autodissociation of the water must be allowed for. The pH of a very dilute acid can never be greater than 7 nor the pH of a base be less than 7.

Buffer Solutions

When studying a chemical reaction which consumes or produces hydrogen ions it is often useful to create conditions under which the pH is nearly constant during the course of the reaction. This can be achieved by the use of a buffer solution, typically a mixture of a weak acid and a salt of that acid with a strong base. If hydrogen ions are produced in the reaction they react with the base to increase the amount of weak acid. If hydrogen ions are consumed then the weak acid dissociates to produce more hydrogen ions. The overall effect is to resist any change in pH. Solutions containing a weak base and its salt with a strong acid have the same effect.

As an illustration of the buffer effect, we will consider a reaction that produces 10^{-3} M HCl. Adding HCl to neutral water to produce this concentration would change the pH from 7 to 3, that is by 4 units. We will now see what happens if 10^{-3} M HCl is produced in a buffer solution containing 10^{-2} M acetic acid and 10^{-2} M sodium acetate.

The acid dissociation constant of acetic acid is 1.75×10^{-5}. If we approximate activities by concentrations then the concentration of ions is determined by the two equations

$$m_{H_3O^+} \cdot m_{Ac^-} = 1.75 \times 10^{-5} m_{HAc}, \tag{10.83}$$

$$m_{H_3O^+} \cdot m_{OH^-} = 1.00 \times 10^{-14}. \tag{10.84}$$

Before the HCl is produced we have the condition of charge balance

$$m_{H_3O^+} + m_{Na^+} = m_{OH^-} + m_{Ac^-}, \tag{10.85}$$

and the mass balances

$$m_{Na^+} = 0.01; \quad m_{HAc} + m_{Ac^-} = 0.02. \tag{10.86}$$

Equations 10.83–10.86 can readily be solved assuming that $m_{H_3O^+}$ and m_{OH^-} are negligible in 10.85 (K_a is small for acetic acid and there is no other source of acid). This assumption leads to

$$m_{Ac^-} = m_{HAc} = 0.01, \quad m_{H_3O^+} = 1.75 \times 10^{-5}, \text{pH} = 4.76. \tag{10.87}$$

After 10^{-3} M HCl is added the charge balance 10.85 is replaced by

$$m_{H_3O^+} + m_{Na^+} = m_{OH^-} + m_{Ac^-} + m_{Cl^-},$$ (10.88)

and we have the additional mass balance

$$m_{Cl^-} = 0.001.$$ (10.89)

These equations can again be solved by neglecting $m_{H_3O^+}$ and m_{OH^-} in 10.88 (for $m_{H_3O^+}$ this may not be obvious as HCl has been added but its validity can be confirmed by the self-consistency of the result). We arrive at the following result

$$m_{Ac^-} = 0.009, \; m_{HAc} = 0.011, \; m_{H_3O^+} = 2.14 \times 10^{-5}, \; pH = 4.67$$

(10.90)

In summary, adding 0.001 M HCl to the buffer solution has moved the HAc/Ac$^-$ equilibrium by 10% and changed the pH by only 0.09 units.

For an effective buffer, that is, one capable of absorbing a reasonably large change in acid concentration, the concentrations of neither component must be too small. The range of buffer activity is therefore limited to about 1 pH unit on either side of the pH for a 50:50 mixture. Acetic acid–acetate buffers can be prepared covering the pH range of about 3.5–5.5 but outside this range other acid–base pairs must be used.

If a weak acid is titrated with a strong base, or a weak base with a strong acid then before neutrality is reached there will be a range over which the pH varies slowly because the mixture is behaving as a buffer. When the end point is reached there will be a sharp change in pH with further addition of

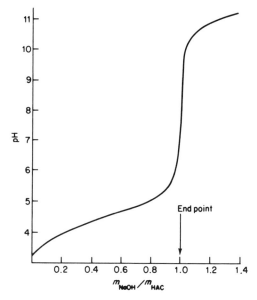

Figure 10.4. Titration curve for 0.10 M acetic acid with NaOH

the titrant. Figure 10.4 shows a typical titration curve for this situation. The ideal indicator in this case will be one that shows a colour change at approximately pH = 8.

Hammett Acidity Function

Concentrated solutions of acids generally show large deviations from ideality. This is particularly true for strong acids, which, when concentrated, often have a much greater ability to protonate weak bases than would be expected from an extrapolation of the properties of their dilute solutions. The activity coefficient $\gamma_\pm$ for an acid like HCl has a concentration-dependence similar to that for a simple salt like LiCl; it decreases up to a concentration of approximately 0.5 M and then increases to values much greater than unity at high concentrations.

The most widely used measure of acidity in concentrated strong acid solutions is that introduced by Hammett and Deyrup and called the Hammett acidity function. If we take the logarithm of 10.74 and use 10.81 as the definition of pH, we have

$$pH = pK_a + \log\left(\frac{a_B}{a_A}\right). \tag{10.91}$$

This is an identity in so far as both pH and pK_a are defined in terms of activities. However, it must be remembered that pK_a is a constant only for dilute solutions because 10.74 is the equilibrium constant for 10.73 only on the assumption that the activity of the water (more generally the solvent) is unity. If K_a' is the exact equilibrium constant for 10.73, allowing for water activity, then the expression analogous to 10.91 is

$$pH = pK_a' + \log\left(\frac{a_B}{a_A a_{H_2O}}\right) \tag{10.92}$$

The Hammett scale of acidity is based upon an experiment in which a small amount of a weak neutral base B is added to the acid solution and the ratio of the concentrations of B and its conjugate acid BH^+ are determined by colorimetric methods. The system B/BH^+ is usually an indicator. From this ratio a quantity H_0 called the acidity function is calculated from the equation

$$H_0 = pK_{BH^+} + \log\left(\frac{m_B}{m_{BH^+}}\right); \tag{10.93}$$

the suffix zero indicates that the function applies only to neutral bases; similar functions for charged bases have been investigated.

By comparing 10.93 with 10.91 we see that H_0 is identical with pH for dilute solutions but in concentrated solutions the two functions differ appreciably. The ratio of activities a_B/a_{BH^+} is very different from the ratio of

240

molalities because the charged species BH^+ deviates much more from ideality than does neutral B. It has, however, been found that the ratio γ_B/γ_{BH^+} is roughly constant for a wide range of organic indicators at a given acid concentration so that H_0 determined from 10.93 is roughly independent of the nature of B.

Figure 10.5 shows the acidity function that has been deduced for sulphuric acid as a function of molality. Very similar graphs are obtained for other very strong acids such as HBr, HCl and $HClO_4$ but weaker acids such as HF and H_3PO_4 give quite different H_0 curves.

It appears to have been well established that the main determining factor for the behaviour of concentrated strong acid solutions is the decrease in water activity. If one assumes that the dominant hydrated proton species in water is $H_9O_4^+$ (described in Chapter 8) then the dependence of H_0 on molality for very strong acids can satisfactorily be explained up to concentrations $m \approx 8 \, \text{mol kg}^{-1}$.[†] However, a cautionary note should be sounded. Although the Hammett acidity function has been shown to be valuable for correlating thermodynamic and kinetic data for organic indicators in concentrated strong acid solutions, there is probably little justification for using it to describe the protonation of other types of weak base.

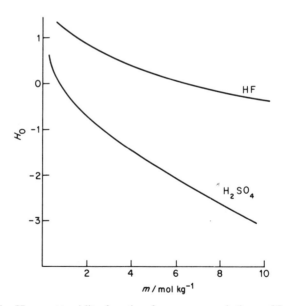

Figure 10.5. The Hammett acidity function for aqueous solutions of H_2SO_4 and HF. (Data from M. A. Paul and F. A. Lang, *Chem. Rev.*, **57**, 1 (1957) and R. P. Bell and K. N. Bascombe, *Disc. Faraday Soc.*, **24**, 162 (1957))

[†]K. N. Bascombe and R. P. Bell, *Disc. Faraday Soc.*, **24**, 158 (1957).

10.5 EQUILIBRIA IN NON-AQUEOUS POLAR SOLVENTS

We have already noted that in water the strong acids such as hydrochloric and perchloric cannot be differentiated in their acid strengths. The reason for this is that water is a relatively good proton acceptor and all of these acids donate, effectively completely, their protons to water molecules. In the Brönsted–Lowry definition of an acid, the strongest acid that is stable in water is H_3O^+; anything stronger would protonate water according to the equation

$$BH + H_2O \rightleftharpoons B^- + H_3O^+, \qquad (10.94)$$

with the equilibrium being established far to the right.

Stronger acids than H_3O^+ are only stable in solvents which are less good proton acceptors than water. For example, in 100% sulfuric acid even perchloric acid has a low dissociation constant, $K_a \approx 10^{-4}$. Very few acids indeed are strong in sulfuric acid as a solvent: $H_2S_3O_{10}$ and $B(HSO_4)_3$ are two with moderately high dissociation constants.

To parallel the above points we note that OH^- is the strongest base that can exist in water; anything stronger would be protonated by water according to the equation

$$B + H_2O \rightleftharpoons BH^+ + OH^-. \qquad (10.95)$$

Ammonia, however, liberates protons much less readily than water and hence liquid ammonia can tolerate a much stronger base. Thus if one needs a very strong base to carry out a chemical reaction, it is likely to be better to use $NaNH_2/NH_3(l)$ rather than $NaOH/H_2O(l)$.

Figure 10.6 shows the ranges over which different solvents can discriminate acid strengths. To take an extreme example, hexane cannot be

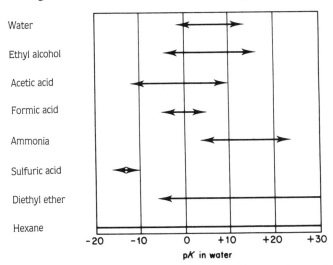

Figure 10.6. Range of existence of acids and bases in different solvents. (From R. P. Bell, *The Proton in Chemistry*, Methuen, London (1959))

detectably protonated or deprotonated hence no acid or base would be dissociated in this solvent and the complete range of acid strengths is, in principle, available (but restricted of course by problems of solubility). Sulfuric acid can discriminate amongst the strong acids but the overall range of discrimination is small because most compounds are protonated in sulfuric acid, that is, most compounds behave as strong bases.

The discrimination range is related to the auto-ionization constant of the solvent. For water we have seen (10.77) that $pK_w = 14.0$. For liquid ammonia auto-ionization is much weaker

$$2NH_3 \rightleftharpoons NH_4^+ + NH_2^-, \quad -\log K = 29.8, \quad (10.96)$$

and for sulfuric acid it is much stronger

$$2H_2SO_4 \rightleftharpoons H_3SO_4^+ + HSO_4^-, \quad -\log K = 3.6. \quad (10.97)$$

Solvents also limit the range of redox strengths that are attainable in solution. However, an important difference from the acid–base situation is that redox reactions are often slow so that an oxidizing or reducing agent that is thermodynamically unstable may nevertheless be kinetically stable.

The strongest oxidizing agent that is thermodynamically stable in water is O_2; anything stronger will liberate O_2 from water. The strongest reducing agent that is stable is, similarly, H_2. Redox strengths are measured by standard electrode potentials. In water under acid conditions we have

$$H^+ + e \rightleftharpoons \tfrac{1}{2}H_2, \quad \mathscr{E}^\ominus = 0 \text{ (definition)},$$

$$\tfrac{1}{4}O_2 + H^+ + e \rightleftharpoons \tfrac{1}{2}H_2O, \quad \mathscr{E}^\ominus = 1.229 \text{ V}. \quad (10.98)$$

Thus reducing agents with negative $\mathscr{E}^\ominus$ values and oxidizing agents with $\mathscr{E}^\ominus > 1.229$ V (for example Ce^{4+}) are unstable in water.

In liquid ammonia, N_2 is the strongest thermodynamically stable oxidizing agent and H_2 the strongest reducing agent. The relevant standard electrode potential is

$$\tfrac{1}{6}N_2 + H^+ + e \rightleftharpoons \tfrac{1}{3}NH_3, \quad \mathscr{E}^\ominus = 0.04 \text{ V}, \quad (10.99)$$

the reference again being zero for the H^+/H_2 couple. A much smaller range of oxidizing agents is thermodynamically stable in liquid ammonia than in water but, as the equilibrium 10.99 is established very slowly, this is not a severe limitation. Solutions of alkali metals in liquid ammonia are powerful reducing solutions even though the $\mathscr{E}^\ominus$ value for say Li^+/Li is negative.

Chapter 11

Solutions of Polymers

11.1 INTRODUCTION

Polymers differ from substances of low molar mass in ways that render it impossible to treat them, as solutes, by the methods described in the earlier chapters of this book. Biological polymers occur widely in Nature but most studies in polymer chemistry are concerned with the synthetic polymers which form the basis of such industries as plastics, packaging films, fibres, adhesives, paints and varnishes. Often, biological polymers are highly complex with respect to the structure of individual macromolecules but simple in that they are composed of macromolecules of uniform size; synthetic polymers, on the other hand, are frequently structurally simple but usually contain molecules covering a wide distribution of molar masses. The breadth of the molecular-weight distribution (MWD) is one of the characteristic features of a synthetic polymer, one of the properties that can be determined quantitatively by the physico-chemical study of solutions of the polymer. (The term 'molecular weight' is equivalent to the term 'molar mass'.)

Chemists would normally regard a molecule as 'very large' if it has a molar mass of, say, $500\,\mathrm{g\,mol^{-1}}$; the molar mass of the exceptionally large molecule of Vitamin B_{12}, for example, is $1330\,\mathrm{g\,mol^{-1}}$. A synthetic polymer would be of no interest if it had a molar mass of less than $10^4\,\mathrm{g\,mol^{-1}}$, indeed values of 10^5 to $10^7\,\mathrm{g\,mol^{-1}}$ are much more common, and it is an easy matter to prepare, for example, polystyrene with a molar mass of $10^8\,\mathrm{g\,mol^{-1}}$. For this reason, there is really no such thing as a dilute solution of a polymer in the sense used earlier in the book where 'dilute' implies that solute-solute interactions can be neglected. Individual polymer molecules are so large that different points *on the same molecule* interact, even if no other polymer molecules are close by. Thus, polymer solutions cannot possibly be ideal *even at infinite dilution* but, paradoxically, certain polymer solutions with finite concentrations obey the laws for ideal solutions quite precisely, as we shall see.

11.2 THE NATURE OF POLYMERS AND POLYMER MOLECULES

The overwhelming majority of synthetic polymers are composed of molecules which contain, linked together, hundreds or thousands of groups of atoms known as the repeating units, which are identified by being obviously derived from the structure of the corresponding monomer(s). Where there is only one type of repeating unit, the substance is called a homopolymer but sometimes two or more types of repeating unit are employed, in which case the substance is a copolymer; for most copolymers, the order of the units along the molecular chain is not regular but statistical, although some ordered arrangements are possible (see Figure 11.1). Some naturally-occuring polymers are much more complex than this, for example, a protein molecule may comprise around twenty types of repeating units in definite sequence. Table 11.1 lists examples of some of the most important types of synthetic polymers, together with one or two of the structurally most simple natural polymers.

Because polymer molecules comprise very long sequences of repeating units (such as A and B, below), the molecular arrangements are often represented by strings of letters, as in Figure 11.1 or simply by lines, as in Figure 11.2. The former shows how binary copolymers—those with two types of repeating units—can be organized in various ways, while the latter illustrates the basic kinds of molecular structures which occur. Cross-linked polymers cannot enter fully into solution but they frequently swell strongly in contact with a liquid which would be a true solvent for the linear or branched polymer counterpart, and measurements of the degree of swelling can yield valuable information about the structure of the polymer.

Vinyl polymers, typified by the structure $+CH_2-CH.X+_n$, constitute a particularly important class because an enormous variety of structures, and hence properties, arises from variation in the nature of the substituent X.

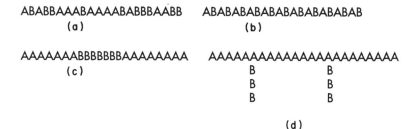

Figure 11.1. Schematic representation of binary copolymers with repeating units A and B: (a) statistical, (b) alternating, (c) block, (d) graft

Table 11.1. Structures of simple polymers

Name	Repeating Unit	Comment
Polyethene (Polyethylene)	$-CH_2-CH_2-$	
Polystyrene	$-CH_2-CH-$ $\vert$ ⬡	
Poly(methyl methacrylate)	$-CH_2-C(CH_3)-$ $\vert$ $COOH_3$	
Polyacrylonitrile	$-CH_2-CH-$ $\vert$ CN	
Poly(vinyl chloride) (PVC)	$-CH_2-CH-$ $\vert$ Cl	
Polyacrylamide	$-CH_2-CH-$ $\vert$ $CONH_2$	Water-soluble
Polybutadiene	$-CH_2-CH = CH-CH_2-$	
Polyisoprene	$-CH_2-C(CH_3) = CH-CH_2-$	

Polyisoprene
(If the substituents at the double bond are in the *cis* arrangement, this is natural rubber)

Name	Repeating Unit	Comment
Polyoxirane	$-CH_2-CH_2-O-$	Water-soluble
Nylon-6,6 (A polyamide)	$-NH(CH_2)_6NHCO(CH_2)_4CO-$	
Poly(ethylene terephthalate) (A polyester)	$-OCH_2CH_2OCO-$⬡$-CO-$	
Cellulose		

Most such polymers dissolve in organic liquids but some are water-soluble; block copolymers may contain one hydrophobic block, e.g., with $X = C_6H_5$, and one hydrophilic block, with $X = OH$. In a simple vinyl homopolymer, it may be possible to ensure a regular sequence of units with respect to the absolute configurations at consecutive chiral (i.e. substituted) backbone carbon atoms. Within a single molecule, the regularity of the sequence may reside in having either identical or strictly alternating configurations, giving rise to structures described as isotactic and syndiotactic, respectively. If the

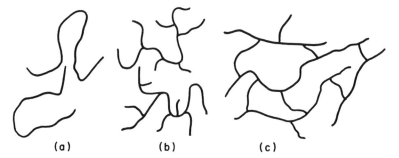

Figure 11.2. Schematic representation of various types of polymer molecules: (a) linear, (b) branched, (c) cross-linked

sequence is irregular, the structure is described as atactic. The arrangement within a group of three consecutive repeating units, known as a triad, can be used to specify the degree of tacticity in the structure, indeed polymer molecules can be characterized quantitatively in terms of the fractional contents of isotactic, syndiotactic and heterotactic triads, shown in Figure 11.3.

The remaining aspect of structure is the breadth of the molecular-weight distribution and the problem of representing a typical polymer by an average molar mass. The principle behind the calculation of an average molar mass is that it should be the molar mass of a uniform (formerly known as 'monodisperse') polymer which has the same value of a specified property as the actual non-uniform (formerly 'polydisperse') polymer—but the difficulty arises that the appropriate average value depends on the particular property chosen as the criterion for this exercise. Thus, the colligative properties (see Chapter 6) depend only on the number of solute molecules present but others depend on both number *and* size; yet others are more complicated in their dependence. Hence, different types of average must be defined, and used appropriately.

Suppose a sample of a polymer contains N_i molecules with i repeating units, each repeating unit having the molar mass M_0. For a colligative

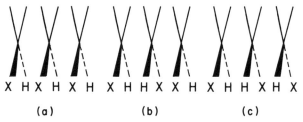

X H X H X H   X H H X X H   X H H X H X

(a)   (b)   (c)

Figure 11.3. Types of triads in vinyl polymers: (a) isotactic, (b) syndiotactic, (c) heterotactic

property, the appropriate average molar mass is then given by equation 11.1, where 'size' is denoted by the index i,

$$M_n = M_0 \sum iN_i \Big/ \sum N_i. \qquad (11.1)$$

For the case where properties are also proportional to size, the corresponding average is given by equation 11.2,

$$M_w = M_0 \sum i^2 N_i \Big/ \sum iN_i. \qquad (11.2)$$

M_n and M_w are known as the number-average and weight-average molar masses, respectively. The number-average is apropriate for use in conjunction with measurements of colligative properties while the weight-average is associated with the measurement of molar mass by means of the scattering of light from polymer solutions; different averages are used in connection with the use of ultra-centifugation and viscosity, as will be mentioned below. The value of the ratio M_w/M_n depends on the form of the whole distribution of molar masses but there are many systems for which it has a value of about 2.

11.3 THE SIZE AND SHAPE OF POLYMER MOLECULES IN SOLUTION

Figure 11.2a represents a polymer molecule as a very thin line of indefinite contour, and it is not difficult to imagine that such a molecule would tend to adopt a relatively highly extended form in contact with a good solvent but that it would curl up tightly on itself if the liquid environment were to interact with it unfavourably. In most solvents, it will probably take on an intermediate, moderately coiled, conformation, effectively occupying a volume in space, which can be characterized by a parameter such as its average radius, however difficult this may be to define. Another way of describing the effective size of a polymer molecule in solution is by means of the time-averaged distance through space between its two end-groups—the root-mean-square (RMS) end-to-end distance.

The simplest theoretical approach to such a molecule is to assume that the constituent atoms are infinitely small, and that bond angles between backbone atoms are infinitely variable. Thus, one imagines the molecule to comprise a sequnce of n bonds, each of length l, absolutely free from any restiction on relative orientation; a statistical calculation can then be performed to assess the RMS end-to-end distance $\langle r^2 \rangle^{1/2}$, which turns out to have the very simple value given by equation 11.3,

$$\langle r^2 \rangle^{1/2} = n^{1/2}l. \qquad (11.3)$$

This basic approach to the theory is known as the random-walk model, and the resulting configuration as a Gaussian chain. The model is unrealistic because it ignores restrictions imposed both by bond angles and the finite size of atoms (collectively called the short-range interactions) and by the fact that elements of space occupied by one part of a polymer molecule are not available for occupation by another part of the same or a different molecule (long-range interactions). The second of these types of limitation is known as the *excluded-volume effect*.

There is a well-known application of the random-walk theory to every-day life. According to the story, some drunkards are clinging to the lamp-post in the very centre of the village square, trying to make the effort to stagger home. As they take steps of equal length but of random orientation, the number of steps each man has to take is proportional the the *square* of the distance from the lamp-post to his house.

More sophisticated treatments go some way towards taking into account the factors neglected by the simplest approach; often the result is that the value of selected measurable property of the polymer solution can be expressed as a virial expansion (see p. 67), in which only the second virial coefficient is important for reasonably dilute solutions. Moreover, when the second virial coefficient is calculated from the theory, it is found to contain a factor of the form $(1 - \theta/T)$, where θ has the dimensions of temperature and is known as the *theta, ideal, or Flory temperature*. The significance of this feature of the theory is that, if measurements are made under the condition $T = \theta$, the solution actually behaves as if it were ideal, because all the virial coefficients can be regarded as being equal to zero. (The theoretical basis for this assertion will be found in Section 11.4.) In these circumstances, *theta conditions* are said to apply and the solvent is described as a *theta solvent*. The dimensions thus found are known as the *unperturbed dimensions* of the macromolecule. Theta conditions can be realized for some polymer/solvent combinations, for example, with polystyrene dissolved in cyclohexane, theta conditions exist at 34 °C and with poly(dimethylsiloxane) in butan-2-one at 20 °C.

If the macromolecule tends to adopt a more extended form than corresponds to theta conditions, the solvent is said to be a *'good' solvent* for the polymer; if the opposite is the case, it is a *'bad' solvent*.

Another parameter used to indicate the size of the space occupied by a polymer molecule in solution is the RMS radius of gyration $\langle s^2 \rangle^{1/2}$. One imagines the macromolecule to be divided up into segments, each with its own centre of mass: the RMS value of the distances of these centres of mass of segments from the centre of mass of the molecule as a whole is the radius of gyration. It is a more fundamental quantity than the RMS end-to-end distance because, according to the theory of macromolecule configurations, it is directly related to the hydrodynamic properties (e.g. viscosity, diffusion), but the two quantities are connected very simply in the case of a Gaussian chain, when equation 11.4 applies,

$$\langle s^2 \rangle = \langle r^2 \rangle / 6. \tag{11.4}$$

To speak of a radius may be thought to imply a sphere but any tendency to think of the macromolecule as approximating in shape to a sphere should be dispelled; after all, one possible extreme configuration is the totally elongated form, with an aspect ratio close to infinity. By calculating the radius of gyration in two directions at right angles, parallel to and perpendicular to the end-to-end vector, it can be shown that, for a Gaussian chain, the parallel RMS radius is exactly double its perpendicular counterpart, so the domain of the polymer molecule rather resembles a rugby or American football with its principal axes in the ratio 2:1. In a good solvent, the aspect ratio can be much higher or, to put it another way, if a system is not under θ conditions, the RMS end-to-end distance is almost certainly larger than that calculated for a Gaussian chain. One way of expressing its size is in terms of the ratio of the actual value $\langle r^2 \rangle^{1/2}$ to that of the ideal value $\langle r^2 \rangle_0^{1/2}$, and to call this ratio the *expansion factor*, α

$$\alpha = \langle r^2 \rangle^{1/2} / \langle r^2 \rangle_0^{1/2}. \tag{11.5}$$

It is commonplace for α to have a value of the order 5–10, and it should be realized that, if this conclusion is true for a linear dimension, the effective volume of the macromolecule (i.e. the volume occupied by the ellipsoidal 'football', mentioned above) may be 125–1000 times greater than calculated from the model. It is unfortunate that the same symbol has been adopted for expansion factor and polarizability.

11.4 THE THEORY OF POLYMER CONFIGURATIONS AND POLYMER SOLUTIONS

The basic thermodynamic theory of polymer solutions was developed independently by M. L. Huggins and P. J. Flory, and is frequently referred to as the Flory-Huggins theory; Flory was awarded the Nobel Prize for Chemistry in 1974.

Enthalpy of Solution

When a polymer is dissolved in a solvent, polymer–polymer interactions and solvent–solvent interactions are partially replaced by polymer–solvent interactions. If we imagine a typical polymer molecule to be divided into sections, known as segments, approximating to the size of a solvent molecule, the enthalpy of mixing ΔH_{mix} can be expressed in terms of the energy change w associated with the formation of a single polymer–solvent interaction by multiplying by the number of such interactions. Calculation leads to the formulation shown in equation 11.6, in which n_1 is the number of moles of solvent, ϕ_2 is the volume fraction of polymer, W is the sum of all the terms w, and Z is a coordination number representing the manner in which the polymer molecule can be thought to be arranged on a lattice. (More about Z in the section on entropy of solution.)

$$\Delta H_{\text{mix}} = ZWn_1\phi_2. \tag{11.6}$$

It is much more convenient to express ZW in terms of units of RT and for this reason we now define the dimensionless parameter χ,

$$\chi = ZW/RT. \tag{11.7}$$

Substituting, we have equation 11.8,

$$\Delta H_{\text{mix}} = RT\chi n_1\phi_2. \tag{11.8}$$

χ is the *polymer-solvent interaction parameter*; shortly, we shall see that it plays a key role in the theory of polymer solutions. (N.B. χ has a relatively low value for a good solvent and high for a poor solvent.)

Entropy of Solution

The entropy change which occurs when a polymer is dissolved is governed by the different number of configurations available to the macromolecule in the dissolved and undissolved states, and the classical way of calculating this difference is by assuming that the structure of the solution can be represented by a cubic lattice, each cell of which contains either a solvent molecule or a segment of the polymer molecule. This assumption can only approach realism if the segment size is very similar to that of the solvent molecule; in the case of vinyl polymers, this often means that the segment is actually a monomer residue.

Figure 11.4a shows a two-dimensional cross-section through such a three-dimensional lattice for a mixture of two compounds with similarly-sized small molecules, e.g. benzene and styrene. The statistical treatment of the different numbers of possible arrangements in the separate and mixed states produces the result that $\Delta S_{\text{mix}} = -R[n_1 \ln x_1 + n_2 \ln x_2]$, where n and x are, respectively, the number of moles and the mole fraction of the species indicated by the subscript. In the case of a polymer solution, e.g. benzene and polystyrene shown in Figure 11.4b, there is the restriction that the segments allocated to a given polymer molecule must run contiguously through the lattice. The lattice has the coordination number Z, mentioned above in connection with the entropy of solution, and it is assumed that a typical polymer molecule is divided into i segments, and so requires i contiguous lattice sites. Application of the parallel statistical reasoning now leads to the result that the entropy of mixing is given by equation 11.9,

$$\Delta S_{\text{mix}} = -R[n_1 \ln \phi_1 + n_2 \ln \phi_2]. \tag{11.9}$$

One sees that this equation is of exactly the same form as in the case of the mixing of small molecules (page 107) but that the volume fraction has replaced the mole fraction.

(a)

(b)

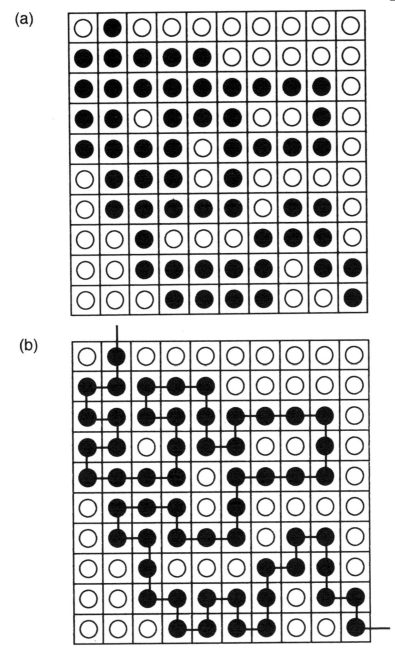

Figure 11.4. (a) Two-dimensional lattice model for a liquid mixture of two substances of similar molecular size, e.g. styrene and benzene. (b) Two-dimensional lattice model of a solution of a polymer in a simple solvent, e.g. polystyrene in benzene. The lattice models provide a basis for the calculation of the difference in entropy of mixing for the two cases (i.e. low molecular weight solute and high molecular weight solute).

Free Energy of Solution

It is a routine matter to combine equations 11.8 and 11.9 in order to obtain the free energy of mixing in equation 11.10,

$$\Delta G_{mix} = RT \left[\chi n_1 \phi_2 + n_1 \ln \phi_1 + n_2 \ln \phi_2\right].$$ (11.10)

Chemical Potential and Activity of Polymer and Solvent

The changes in chemical potential of the solvent and the polymer on mixing are, respectively, $(\mu_1 - \mu_1^0)$ and $(\mu_2 - \mu_2^0)$; from equation 6.34, it is evident that they can be evaluated by differentiating equation 11.10 with respect to n_1 or n_2, as appropriate, remembering the relations $\phi_1 = n_1/(n_1 + in_2)$ and $\phi_2 = in_2/(n_1 + in_2)$. The results are best expressed in the forms of equations 11.11 and 11.12,

$$(\mu_1 - \mu_1^0) = RT[\chi \phi_2^2 + \ln(1 - \phi_2) + (1 - 1/i)\phi_2],$$ (11.11)

$$(\mu_1 - \mu_2^0) = RT[\chi i \phi_2^2 + \ln \phi_2 + (1 - i)(1 - \phi_2).$$ (11.12)

In ideal situations, $(\mu_1 - \mu_1^0)$ is equal to $RT \ln x_1$ but when there is deviation from ideality, it is $RT \ln a_1$, where a_1 is the activity of species 1 (the solvent). The relationship between the activity of the solvent and the composition of the polymer solution is thus, according to the lattice theory, given by equation 11.13,

$$\ln a_1 = \chi \phi_2^2 + \ln(1 - \phi_2) + (1 - 1/i)\phi_2.$$ (11.13)

The term $\ln(1 - \phi_2)$ can be expanded as a convergent series, according to the general formulation $\ln(1 - x) = -x - x^2/2 - x^3/3 - x^4/4 - \ldots$, which, after truncation between the second and third terms, reduces equation 11.13 to equation 11.14.

$$\ln a_1 = -\phi_2/i - (1/2 - \chi)\phi_2^2$$ (11.14)

If conditions can be found where the polymer–solvent interaction parameter χ is exactly equal to $1/2$, the second term on the right-hand side of equation 11.14 vanishes and ideal behaviour would be observed. This second term can be regarded as an *excess chemical potential*, i.e. a correction term accounting for non-ideality, given by equation 11.15,

$$(\mu_1 - \mu_1^0)^E = -RT(1/2 - \chi)\phi_2^2.$$ (11.15)

As this is a free energy expression, it can be broken down into enthalpic and entropic components, and it has been shown that the enthalpic contribution can take the form $\Delta H_1 = RT\kappa_1 \phi_2^2$, where κ_1 is a constant. It therefore becomes tempting to formulate the entropy term as $\Delta S_1 = R\psi_1 \phi_2^2$, where ψ_1 is another constant, and so arrive at equation 11.16,

$$(\mu_1 - \mu_1^0)^E = -RT(\kappa_1 - \psi_1)\phi_2^2.$$ (11.16)

The two new constants, κ_1 and ψ_1, have the same dimensions, so the constant θ, with the unit of temperature, can be defined by equation 11.17, and equation 11.16 converted into equation 11.18,

$$\theta/T = \kappa_1/\psi_1, \qquad (11.17)$$

$$(\mu_1 - \mu_1^0)^E = -RT\psi_1(1 - \theta/T)\theta_2^2. \qquad (11.18)$$

This is the theoretical basis for the conclusions presented on page 248. One sees that $\chi = 1/2$ and $T = \theta$ are equivalent statements of theta conditions.

A further development of the theory provided the equation for the expansion factor, given in equation 11.19,

$$\alpha^5 - \alpha^3 = 2C\psi_1(1 - \theta/T)M^{1/2}. \qquad (11.19)$$

In this equation, C is a complex constant for the system and M is the molar mass of the polymer.

11.5   EXPERIMENTAL METHODS FOR THE STUDY OF POLYMER SOLUTIONS TO DETERMINE MOLAR MASSES

Osmotic Pressure

If a pure solvent and a solution of a chosen solute in the same solvent are separated by a membrane which allows the solvent but not the solute to diffuse through freely, the solvent will tend to pass through the membrane in order to reduce the chemical potential of the solute by dilution. The geometry of the containing vessels can be designed so that the result is a build-up of hydrostatic pressure on the solution side of the membrane to the point where further passage of solvent is prevented and an equilibrium is established: the excess hydrostatic pressure on the solution side is then the osmotic pressure π of the solution at the prevailing concentration c of the solute. The general relation between these two quantities is commonly represented by a virial expansion in the form of equation 11.20,

$$\pi/c = RT(A + Bc + Cc^2 + ...). \qquad (11.20)$$

Figure 11.5 shows the normal type of curve obtained in practice; clearly, it is difficult to extrapolate measurements made at finite values of c back to $c = 0$. The line shown in the figure for experiments under theta conditions (where $B = C = 0$) demonstrates the simplicity of interpretation in such cases. For polymer solutions, the value of osmotic pressure measurements resides in the theoretical identification of the first virial coefficient with the reciprocal of the number-average molar mass, as in equation 11.21,

$$A = 1/M_n. \qquad (11.21)$$

The second virial coefficient represents the extent of deviation from ideality arising from interactions between the polymer molecules in the solution and, more specifically, it provides an estimate of the magnitude of the excluded

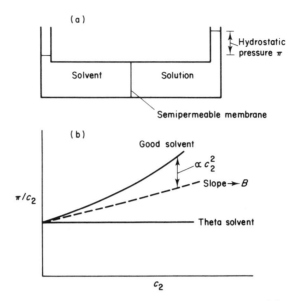

Figure 11.5. (a) Representation of osmotic pressure apparatus. (b) Dependence of π/c_2 on c_2 for polymer in a theta solvent and in a good solvent

volume effect. If v is the excluded volume, and $\mathcal{N}$ is Avogadro's number, the theory developed above leads to the value of the second virial coefficient B given in equation 11.22,

$$B = \mathcal{N}v/2M_n \qquad (11.22)$$

The theory also leads to the simple expression for the third virial coefficient C, given in equation 11.23,

$$C = gB^2/M_n. \qquad (11.23)$$

In this expression, g is a constant with the value 0.5 for the case of Gaussian chains in a good solvent.

Light Scattering

Any interaction between electromagnetic radiation and matter results in some of the radiation being scattered, i.e. it emerges from the encounter travelling in a direction different from that of the incident beam. Light is scattered from polymer molecules in solution (unless the polymer in solution and the solvent have precisely the same refractive index), and quantitative measurements of this scattering can provide detailed information about the size, shape and dynamic behaviour of the polymer molecules. As with osmotic pressure measurement, the method is absolute; no assumptions are involved and no reference material is necessary. However, it must be noted that polymer molecules can fall into either of two size regimes; it is necessary to make separate treatments of 'small' and 'large' scattering particles, the size being judged in comparison with the wave-length λ of the incident light. 'Small' implies that the diameter of the solute particle (molecule, in the present case) is certainly not more than $\lambda/10$ and preferably less than $\lambda/20$. For obvious practical reasons, practitioners of this method prefer to use visible light, and it follows that only those polymers with molar masses up to about 2×10^5 g mol^{-1} can be regarded as small.

The intensity of light scattered is proportional to the number of scattering particles per unit volume, but it is also dependent on the molar mass of the scattering particle, 'heavier' molecules scattering more efficiently than 'lighter' ones; for this reason, the appropriate mean molar mass in this case is the weight-average.

If monochromatic light is employed, scattering is accompanied by a slight broadening of the wave-length distribution. This phenomenon, *quasi-elastic light-scattering*, is a valuable but separate subject which will not be considered further here. The Raman effect also occurs alongside light-scattering but its contribution is always neglected.

Figure 11.6 shows a schematic representation of the experimental set-up. The essential parameters are: the incident intensity of the light beam I_0, the emergent intensity I_θ at distance r at angle θ to the direction of incident beam, the number of scattering particles per unit volume N_v, and the polarizability α of the scattering material (see Chapter 8). Rayleigh showed that the relation between these quantities is provided by equation 11.24,

$$I_\theta = 8\pi^4 \alpha^2 I_0 \, (1 + \cos^2\theta) N_v / r^2 \lambda^4 \tag{11.24}$$

This expression is usually re-arranged in the form of equation 11.25, where the intensities and the geometric factors are amalgamated into the composite entity known as the Rayleigh ratio R_θ,

$$R_\theta = I_\theta r^2 / (1 + \cos^2\theta) I_0 = 8\pi^4 \alpha^2 N_v / \lambda^4. \tag{11.25}$$

The polarizability α is not a convenient quantity to measure work with, but it is closely related to the weight-average molar mass and the refractive

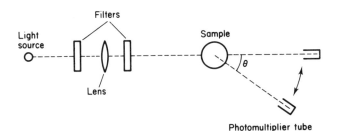

Figure 11.6. Schematic reperesentation of apparatus for measuring light scattering as a function of angle θ

index n or rather the gradient of refractive index with respect to solute concentration, dn/dc, according to equation 11.26,

$$\alpha = (M_w/2\pi\mathcal{N})/(dn/dc). \tag{11.26}$$

It is therefore necessary to determine dn/dc for any newly-investigatd system, as well as to carry out the essential measurements of intensity of the scattered light. Using the substitution in equation 11.27, where c is the concentration and V is the volume,

$$N_v = c\,\mathcal{N}/VM_w; \tag{11.27}$$

combining equations 11.25 and 11.26, and gathering various constants into the term K, as in equation 11.28,

$$K = 2\pi^2 n_0^2/\lambda^4 \mathcal{N}(dn/dc)^2; \tag{11.28}$$

the final expression obtained is equation 11.29,

$$R_\theta = KcM_w. \tag{11.29}$$

This equation is an over-simplification. Firstly, it is only true in the limit of infinite dilution; secondly, it is only true in the limit of scattering at zero angle. Since measurements are made at finite concentrations and finite scattering angles, it is necessary to extrapolate the observations in respect of both parameters (concentration and angle), and an ingenious method of carrying out this requirement is the Zimm plot, in which values of Kc/R_θ are plotted against $[\sin^2(\theta/2) + kc]$, where k is an arbitrary number, often 100 or 1000. Figure 11.7 demonstrates the use of the Zimm plot to make a double extrapolation to the ordinate axis, and so to identify the value of M_w.

In practice, the light-scattering apparatus measures I_θ for known values of

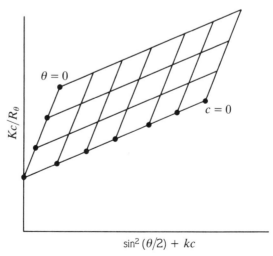

Figure 11.7. Representation of a Zimm plot according to equation 11.11

the instrument parameters, hence R_θ is evaluated for various c and θ. The results can be transferred to a computer, programmed to draw the Zimm plot and to report the doubly-extrapolated value of M_w.

The effect of concentration can, as usual, be expressed in the form of a virial equation, as in equation 11.20.

$$Kc/R_\theta = (M_w)^{-1} + 2Bc + 3Cc^2 + \ldots \qquad (11.30)$$

The initial slope of the concentration-dependence of the left-hand side of this equation is clearly a measure of the second virial coefficient, and it can be shown that the dependence on $\sin^2(\theta/2)$ yields the value of the RMS radius of gyration. Both of these quantities can be evaluated by analysis of the Zimm plot (by the computer, if available).

Viscosity of Dilute Solutions

By far the simplest technique for the quantitative study of polymer solutions is viscosity measurement. With the most elementary of equipment, precise determinations of flow times can be made, but unfortunately it is a complex matter to extract from the results reliable estimates of molar mass; for comparison of average molar masses, the technique is excellent.

Even a very small amount (say, 0.1% w/w) of a polymer of high molar mass significantly raises the viscosity of a liquid in which it is dissolved. Viscosity is measured by monitoring the flow of a given quantity of the liquid through a capillary tube under a hydrostatic head in a simple viscometer, usually essentially of the type introduced by Ostwald and shown in Figure 11.8. The basic measurements are the flow times of the solvent and the solution, t and t_o, respectively. If the corresponding viscosities, η and η_0, are proportional to the flow times, the fractional increase in viscosity arising from the presence of the polymer (the *relative viscosity increment*) is $(\eta - \eta_0)/\eta_0$, and allowance for the concentration of polymer producing this effect can be made by dividing by c to give what is known as the *reduced*

Figure 11.8. Ubbelohde suspended level dilution viscometer

viscosity or *viscosity number*. Extrapolation of this quantity to infinite dilution (to eliminate the effects of polymer–polymer interaction) gives the *limiting viscosity number* or *intrinsic viscosity*, $[\eta]$. Empirically, the relation connecting the intrinsic viscosty with the molar mass takes the form of equation 11.31, often known as the Mark–Houwink–Kuhn–Sakurada equation. M_v has a value intermediate between those of M_n and M_w,

$$[\eta] = KM_v^a. \tag{11.31}$$

Here, K and a are constants for the combination of polymer and solvent at the given temperature, and M_v is the molar mass, in this case the viscosity average, mentioned on page 246. To be really useful, it is necessary to use an absolute technique, such as light-scattering, to calibrate the equation.

The usual theory of polymer solutions regards the solute molecule as something like a pearl necklace, a linear series of connected beads, but for viscosity there are two important limiting versions of the model. In the first, the beads are sufficiently far apart that the disturbance to flow caused by one bead is not felt by its neighbours; this model is known as the freely-draining coil. The alternative assumes that the disturbances from neighbouring beads overlap so much that the solvent is effectively trapped inside the polymer molecule. Thus, the polymer and entrapped solvent move as a single entity, rather like a solid sphere, with a radius equal to the mean end-to-end distance of the polymer molecule; such a sphere is said to be the macromolecule's *equivalent hydrodynamic sphere*. The latter concept leads to equation 11.32 where Φ is, in principle, a universal constant for all flexibly coiled polymers,

$$[\eta] = \Phi \langle r^2 \rangle^{3/2} / M_v. \tag{11.32}$$

There is a basic assumption in the theory that the segments of the polymer molecule are distributed in space in Gaussian fashion, as would be expected at the θ temperature, but more generally equation 11.33 applies, where α is here the expansion factor,

$$[\eta] = \alpha^3 \Phi \langle r^2 \rangle^{3/2} / M_v \tag{11.33}$$

Comparison of measurements at an arbitrary temperature T and at $T = \theta$ provides a value for α at temperature T because $\alpha_T = ([\eta]_T/[\eta]_\theta)^{1/3}$. More sophisticated theories find the exponent in this equation to be 0.41 rather than $\frac{1}{3}$.

Ultracentrifugation

It is well-known that a heavier solid can be separated from a lighter liquid by centrifugation, but it is not so widely appreciated that, if a heavier solute dissolved in a lighter solvent is subjected to a centrifugal potential, the solute molecules will tend to move in the direction away from the axis of rotation, that is, it will undergo sedimentation. Potentially, this fact provides a basis for the determination of the molar mass of the solute but it requires a strong field—up to some 500 000 times the force due to gravity; this can be achieved in the ultracentrifuge, shown diagrammatically in Figure 11.9.

As sedimentaion proceeds, the concentration of polymer drops at the inner (narrow) end of the wedge-shaped cell and increases at the outer (broad) end. The centrifugal field produces a concentration gradient, which is opposed by diffusion attempting to maintain a uniform concentration of solute, and the two competing tendencies may result in an equilibrium. A (more-or-less diffuse) boundary develops between pure solvent at the inner

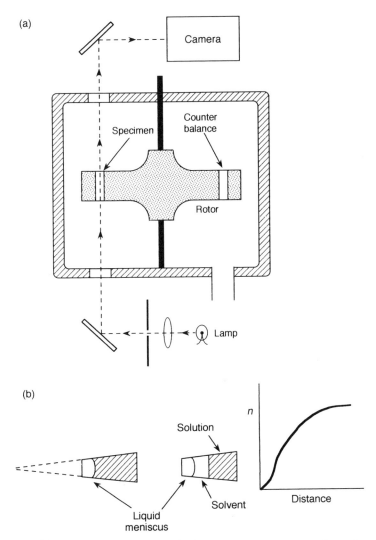

Figure 11.9. (a) Vertical section through the rotor compartment of an ultracentrifuge. (b) Horizontal section through the specimen cell, before and after centrifugation, together with typical plot of refractive index versus distance from the rotor axis

end and solution further out: the determination of either the position or the rate of movement of this boundary provides the necessary data for evaluation of the molar mass of the solute. Various optical techniques have been used to observe the boundary, and refractive index is a popular property for this purpose. There are two general approaches to the determination of molar mass, using different levels of centrifugal field.

The *sedimentation equilibrium* method employs a relatively low field, in the range 10,000–20,000 r.p.m, so that an equilibrium is set up. This provides a situation that can be treated thermodynamically but it takes a long time to establish the necessary equilibrium. The observation of the complete profile of concentration *versus* distance moved by the solute enables both M_w and M_z $(= M_o \Sigma i^3 N_i / \Sigma i^2 N_i)$ to be evaluated.

The *sedimentation velocity* method is carried out at high centrifugal field, and operates in such a way that diffusion is completely negligible, and one simply observes the hydrodynamic character of the system. In the case of a polymer with a significantly broad distribution of molecular weights, the boundary between the solvent and the solution will not be sharp because solute molecules with different molar masses will move with different velocities; it is, in principle, possible to make use of this fact to estimate the breadth of the distribution but it is a difficult task. Usually, one is content to deduce the average molar mass, however, it is necessary to understand which particular average is obtained. This actually depends on details of the procedure adopted but it is usually M_w, or something close to it.

The ultracentrifuge is a powerful tool because it covers the widest molar-mass range of all methods, 10^2–10^6 g mol^{-1} but it is very expensive. A small number of laboratories use the technique routinely for synthetic polymers but it is more widely employed for biological polymers which have more nearly uniform molar masses. Svedberg was awarded the Nobel prize for Chemistry in 1926 for developing the technique.

Size-exclusion Chromatography

In this technique, formerly known as Gel-Permeation Chromatography, a sample of a solution of polymer (typically, 1 ml of a 0.1% solution) is added to the top of a column packed with very fine, highly porous, beads, and pure solvent is added so as to follow the solution through the column, (with pumping, if necessary). As the solution travels down the column, the solute molecules have an opportunity to diffuse into the pores in the beads but any macromolecules which are too large to enter the pores will simply flow past the packing, and be eluted rapidly at the end of the column. Smaller solute molecules will enter the pores, and continue down the column only after diffusing out again into the stream of solvent. The speed with which the solute molecules emerge from the beads varies according to their effective dimensions, i.e their hydrodynamic volumes, the smallest molecules being eluted last. The precise mechanism of the process is at present a matter of dispute, but it shoud be realised that the polymer molecules can be separated according to size, even when the beads contain a narrow distribution of pore sizes.

Thus, the eluted liquid contains the polymer molecules distributed according to size, i.e. fractionation has occurred (See Section 11.5.); a suitable type of detector can record the amount of polymer molecules in each size band, and thus present a picture of the molecular weight distribution, together with the number-average and weight-average molar masses. Interpretation of the data so obtained is not necessarily straightforward; ideally, the system requires calibration with samples of the same polymer of known molar mass but these are usually not available.

Calibrated polystyrene samples are frequently used to set up a so-called universal calibration, based on the experimental observation that the product of intrinsic viscosity and molar mass, $[\eta]M$, is the same function of the elution volume (the volume of solvent passed through the column up to that moment) for a wide range of polymers. If this generalization is valid, the value of $[\eta]M$ for a known polymer (denoted by suffix '1') and a sample polymer (suffix '2') must be the same at points for the same elution volume, i.e. for a given elution volume, $[\eta]_1 M_1 = [\eta]_2 M_2$. Substituting from equation 11.31, one finds

$$\log M_2 = \frac{1}{(1 + a_2)} \log \frac{K_1}{K_2} + \frac{(1 + a_1)}{(1 + a_2)} \log M_1,$$

and the molar mass of the sample can easily be calculated if the Mark–Houwink–Kuhn–Sakurada viscosity constants (see page 258) are known for *both* polymers. The use of the parameter $[\eta]M$ can be justified theoretically; in fact, it represents the hydrodynamic volume of the macromolecule.

Inverse Gas Chromatography

This technique was mentioned briefly on page 158. It provides a basis for measuring the interaction parameter for a polymer–solvent system at both finite and infinite dilution. It also facilitates the determination of the solubility parameters of polymers, and it is especially useful for studying the glass transition and other phase transitions in solid polymers, including those involving liquid-crystalline states.

11.6 PHASE BEHAVIOUR

A polymer and a good solvent will probably be miscible in all proportions but with a 'less good' solvent the same may not be true in certain temperature regimes. A polymer solution may exhibit an upper or lower critical solution temperature (consolute temperature) or both. In a region of partial miscibility, the system separates into a polymer-rich and a polymer-poor phase, and such separation is much more common than with solutes of low molar mass; this results from the fact that the entropy gained by a polymer molecule on entering a solution is much lower because the segments inevitably remain bound together; hence, it is much more likely that opposing enthalpic and entropic factors will fall into balance. A corollary of this line of reasoning is that the temperature at which phase-separation will occur is a function of the size of the polymer molecule, higher molar mass components tending to separate before those of lower molar mass. It can be shown that the temperature at which polymer of infinite molar mass just becomes insoluble is the theta temperature.

A process of gradually changing conditions so as progressively to precipitate molecules of lower and lower size is known as *fractionation*. It can be achieved by reducing the temperature or by adding to the system a non-solvent for the polymer. This brings about a progressive increase in χ until it reaches the critical value of $\frac{1}{2}$ (see equation 11.15). Polymer fractions are useful for the calibration of non-absolute methods of molar mass determination because they can be regarded as comprising molecules of uniform size.

11.7 READING LIST

The object of this chapter is to highlight the special properties of solutions of polymers in relation to the treatment of solutions in general in the earlier chapters. As a discussion of polymer solutions, it constitutes merely an outline of the subject, which some readers may wish to explore in greater depth. Reference is therefore made to the excellent text *Polymers: Chemistry & Physics of Modern Materials* by J. M. G. Cowie, (Blackie), Second Edition (1991), and to the classical work *Principles of Polymer Chemistry* by P. J. Flory, (Cornell University Press) (1953).

Chapter 12

Liquid Interfaces and Adsorption Phenomena

12.1 INTRODUCTION

A liquid can form three types of interface: (1) with its vapour or a mixture of its vapour and another gas; (2) with another liquid with which it is immiscible or only partially miscible; and (3) with a solid. There are two other types of interface, solid–gas and solid–solid. The science of interfaces is concerned with the nature of the boundary region between phases and the consequences of its existence. The word surface as applied to solids and liquids is generalized to interface. Systems involving interfaces are broadly classified into those with phases of macroscopic extent, and into a separate class—called colloids—when at least one of the dimensions of a phase is in the submicrometer region usually implying the presence of a microscopic dispersed phase. This second class comprises the colloids, dealt with separately in Chapter 13. Confining the discussion here to interfaces with at least one liquid, it will be appreciated that there are many facets to the subject because the liquids can, for instance, be solutions of electrolytes or non-electrolytes and can include synthetic and biological macromolecules.

Interfacial and colloidal phenomena play important roles in animate and inanimate natural systems and in man-made systems: the behaviour of natural water surfaces (e.g., lakes and water in soils); the structure and physical and chemical function of biological systems; a variety of analytical techniques such as liquid–solid and gas–liquid chromatography; numerous industrial processes and products; and many households and pharmaceutical commodities rely strongly on them.

When the surface of a liquid is extended reversibly by area $d\mathcal{A}$ at constant temperature, volume and chemical composition, the change in the Helmholtz free energy of the system is

$$dA = \left(\frac{\partial A}{\partial \mathcal{A}}\right)_{T,V,n_i} d\mathcal{A}. \tag{12.1}$$

The partial differential coefficient in 12.1 is one definition of the surface tension of the liquid:

$$\gamma = (\partial A / \partial \mathscr{A})_{T,V,n_i}. \tag{12.2}$$

The surface tension is the force per unit length acting in the liquid surface. (The hitherto common unit dyne cm^{-1} for surface tension has been replaced by the SI unit $mN\,m^{-1}$, its numerical equivalent.)[†] Table 12.1 gives a few representative values of surface and liquid–liquid interfacial tensions. The value of the surface tension of water is greater than for other simple liquids, values for which fall in the range 20–40 $mN\,m^{-1}$; values for liquid metals are exceptionally high. Methods of measuring γ can be classified as static, involving the relation of interface shape to the shape for a dimensionlesss model system, e.g., the pendent-drop method, or as dynamic, such as the determination of the volume of the detaching portion of a pendent drop (drop-volume method). The underlying principles are embraced in a branch of interfacial science known as capillarity which we examine in section 12.3. The reader is referred to standard textbooks for methods of interfacial tension determination.[‡] When one phase is a solid it is difficult to determine the analogous solid–solid interfacial tensions and indirect methods are used.

The concept of surface tension can also be introduced from microscopic or statistical mechanical theory, since it originates from intermolecular attraction.[§] One advantage of the microscopic approach is that it enables computer modelling of some interfacial properties to be carried out which would be difficult to measure experimentally.[‖] Elegant treatments exist to account for the existance of surface energy but it is instructive to use a simple approach. A typical molecule embedded in the bulk liquid requires its share of the molar enthalpy (and free energy) of vaporization to lose its neighbouring 'liquid' contacts and escape to the vapour. To create new

Table 12.1. Surface and interfacial tensions

	$\gamma/mN\,m^{-1}$		$\gamma/mN\,m^{-1}$
Water	72.8	Acetone	23.7
Ethanol	22.3	Propylene carbonate	41
Benzene†	28.9	Liquid argon (at 90 K)	12
Carbon tetrachloride	26.8	Mercury	485
Water/carbon tetrachloride	45	Water/hexane	51
Water/mercury	375		

†See also Table 12.3.

---

†The symbols and nomenclature in this chapter closely follow IUPAC recommendations (*Pure Appl. Chem.*, **31**, 579 (1972)). It is appropriate in this section to use the Helmholtz free energy $A = U - TS$ rather than the Gibbs function. Phases are denoted by superscripts.
‡e.g. A. W. Adamson, *Physical Chemistry of Surfaces*, 3rd edn, Interscience, Chichester (1977).
§G. Navescues, *Report. Progr. Phys.*, **72**, 1131 (1979).
‖J. S. Rowlinson, *Chem. Soc. Rev.* (1978).

liquid surface requires the loss by each new surface molecule of about one-quarter of the neighbouring contacts (say, 3 out of 12) it had in the bulk liquid. On this basis the surface energy should be approximately a quarter of the energy of vaporization.

12.2 ADSORPTION AT INTERFACES

When at least one of the phases forming an interface is a solution, the composition of the system in the vicinity of the interface is expected to differ from that of the bulk. One component, at least, will usually have an enhanced concentration and be said to be adsorbed. The interfacial region, whose thickness is at most a few molecular diameters, might intuitively be regarded as being a physically heterogeneous layer sandwiched between the bulk phases, e.g. between AB and CD in Figure 12.1. To quantify the amount adsorbed, reference is made to a hypothetical system of the same volume as the real system. The surface excess amount or the Gibbs absorption, denoted n_i^σ, of component i is defined as the excess amount of this component actually present over that present in the reference system in which the bulk concentrations of the two phases remain uniform up to the presumed plane of discontinuity between the phases, which is called the Gibbs dividing surface. Accordingly,

$$n_i^\sigma = n_i - V^\alpha c_i^\alpha - V^\beta c_i^\beta, \tag{12.3}$$

where n_i is the total amount of component i in the system, c_i^α and c_i^β are its concentrations in the two bulk phases α and β, and V^α and V^β are the volumes of the phases defined by the Gibbs dividing surface. There is no volume specially assigned to an adsorbed substance in this description.

Figure 12.2 shows how the concentration c_i in an actual system might vary in the z-direction of Figure 12.1, and it also shows the Gibbs dividing surface positioned between the surfaces which are in the bulk phases. In the vicinity of the Gibbs surface, c_i^α and c_i^β do not exist as bulk values, but there is a variable concentration c_i depending on the position in the z direction of the volume element dV:

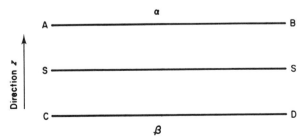

Figure 12.1. Representation of the surface region where bulk phase concentrations exist above AB and below CD: the Gibbs surface is SS

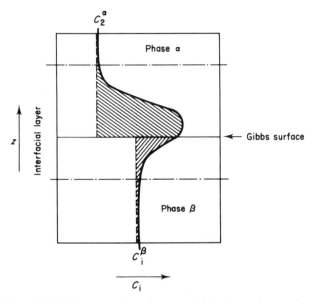

Figure 12.2. Concentration changes within the surface region

$$n_i^\sigma = \int_\alpha (c_i - c_i^\alpha)\,\mathrm{d}V + \int_\beta (c_i - c_i^\beta)\,\mathrm{d}V. \qquad (12.4)$$

If the surface area $\mathscr{A}$ is known, the Gibbs surface concentration or the surface excess concentration Γ_i^σ is given by

$$\Gamma_i^\sigma = n_i^\sigma/\mathscr{A}. \qquad (12.5)$$

Corresponding definitions can be given in terms of numbers of molecules or of mass.

In general, the choice of position of a Gibbs surface is arbitrary but the amounts of the components in a reference system depend on its position. Quantities can be obtained which do not depend on its precise location, and their specification forms the basis for experimental measurements which cannot take the arbitrary nature of the Gibbs surface into account. For a multicomponent system the surface excess concentrations of components 1 and i, Γ_1 and Γ_i, respectively, for an arbitrarily chosen Gibbs surface lead to the relative adsorption $\Gamma_i^{(1)}$ of component i with respect to component 1,

$$\Gamma_i^{(1)} = \Gamma_i^\sigma - \Gamma_1^\sigma \left(\frac{c_i^\alpha - c_i^\beta}{c_1^\alpha - c_1^\beta} \right), \qquad (12.6)$$

which corresponds to choosing the Gibbs surface so that $\Gamma_1^\sigma = 0$ and $\Gamma_1^{(1)} = 0$ in the reference system.

For liquid–vapour interfaces when the vapour pressure is low, a useful approximation is

$$\Gamma_i^{(1)} = (n_i - n_1 x_i / x_1) / \mathscr{A}, \tag{12.7}$$

where x_1 and x_i are the mole fractions of components 1 and i in the liquid phase.

This description of an interface can be extended to give a system of surface thermodynamics which takes surface areas, interfacial tensions and adsorption phenomena into account. Important relationships can be obtained by this method. An energy U^σ and an entropy S^σ are assigned to the surface to make the model system equivalent in these respects to the real system,

$$U = U^\sigma + U^\alpha + U^\beta; \quad S = S^\sigma + S^\alpha + S^\beta. \tag{12.8}$$

The Helmholtz free energy of the system is

$$A = A^\sigma + A^\alpha + A^\beta; \quad A^\sigma = U^\sigma - TS^\sigma. \tag{12.9}$$

Quantities can be referred to unit surface area, e.g., the surface free energy per unit area is

$$a^\sigma = A^\sigma / \mathscr{A}. \tag{12.10}$$

The total differential of the system is

$$dA = -S\,dT - P^\alpha\,dV^\alpha - P^\beta\,dV^\beta + \gamma\,d\mathscr{A}$$
$$+ \sum_i \bar{\mu}_i^\alpha\,dn_i^\alpha + \sum_i \bar{\mu}_i^\beta\,dn_i^\beta + \sum_i \mu_i^\sigma\,dn_i^\sigma, \tag{12.11}$$

involving obvious bulk terms, but the chemical potentials $\bar{\mu}_i^\alpha$, $\bar{\mu}_i^\beta$ and μ_i^σ must be defined. Partial differentiation of equation 12.9, subject to the usual constraints, gives

$$\bar{\mu}_i^\alpha = \mu_i^\alpha + (\partial A^\sigma / \partial n_i^\alpha)_{n_j^\alpha, n_i^\beta, n_i^\sigma}, \tag{12.12}$$

with an analogous equation for $\bar{\mu}_i^\beta$, where μ_i^α is the usual bulk chemical potential defined in terms of the Helmholtz rather than the Gibbs free energy,[†] and, at constant surface area,

$$\mu_i^\sigma = \partial A^\sigma / \partial n_i^\sigma. \tag{12.13}$$

The second term on the right-hand side of equation 12.12 arises because the surface is not strictly autonomous; its properties depend on variables pertaining to phases α and β. In equilibrium studies this second term can be neglected and autonomy assumed, but the power of surface thermodynamics to apply to certain non-equilibrium situations should be noted.

Using standard thermodynamic formulae it can be shown that

$$dA^\sigma = -S^\sigma\,dT + \gamma\,d\mathscr{A} + \sum_i \mu_i^\sigma\,dn_i^\sigma. \tag{12.14}$$

---

†Chemical potentials of the form $\partial A / \partial n_i = \mu_i$ can be defined involving the Helmholtz function; in the context of interfaces the set of chemical potentials is slightly more complicated than for bulk phases alone.

Appropriate integration gives

$$A^\sigma = \gamma \mathscr{A} + \sum_i \mu_i^\sigma n_i^\sigma, \qquad (12.15)$$

which, recalling that $\Gamma_i^\sigma = n_i^\sigma/\mathscr{A}$, yields

$$\gamma = a^\sigma - \sum_i \Gamma_i^\sigma \mu_i^\sigma. \qquad (12.16)$$

Clearly the interfacial tension γ and the surface free energy per unit area are not identical quantities. Indeed, γ is experimentally measurable, independently of any choice of Gibbs surface, whereas the value of a^σ depends on the position of the dividing surface.

When equation 12.15 is differentiated and compared with equation 12.14 under conditions of constant temperature,

$$d\gamma = -\sum_i \Gamma_i \, d\mu_i, \qquad (12.17)$$

is the adsorption isotherm often referred to as the Gibbs equation. It is the fundamental equation for relating the interfacial tension to the amount adsorbed when $d\mu_i$ is suitably expressed in experimentally measurable terms. The use of the Gibbs adsorption equation will now be illustrated.

Adsorption at Liquid–Vapour and Liquid–Liquid Interfaces

In many cases the adsorption of one component of a solution at a liquid–vapour or at a liquid–liquid interface (the latter type is much less studied) gives a layer one molecule thick, called a soluble monolayer. Certain insoluble substances can be made to spread at a liquid surface (usually aqueous) giving insoluble monolayers which are discussed in Section 12.3. Appropriate forms of the Gibbs adsorption equation can be used to obtain the amount of a soluble substance adsorbed.

For solute component 2 in solvent 1, the chemical potential of 2 can be written as a function of its activity,

$$\mu_2^1 = \mu_2^{\ominus,1} + RT \ln a_2^1. \qquad (12.18)$$

The relative adsorption, equation 12.6, of component 2 with respect to 1, $\Gamma_2^{(1)}$, corresponds to putting $\Gamma_1 = 0$ in equation 12.17 which becomes

$$d\gamma = -\Gamma_2^{(1)} RT \ln a_2^1. \qquad (12.19)$$

Equation 12.19 is applicable to a component which does not dissociate in solution, e.g., a simple alcohol in water. The adsorption at a liquid–liquid interface can similarly be obtained. For an ideal or sufficiently dilute solution the concentration c_2 can replace a_2^1, and the slope of experimentally determined γ against c_2 plots gives $\Gamma_2^{(1)}$.

There is considerable interest in amphipathic molecules i.e. molecules having a non-polar hydrocarbon and a polar hydrophilic portion, which are known as surfactants (including soaps and detergents). They may be anionic, cationic or non-ionic, and will be discussed in more detail in Chapter 13. The anionic surfactant sodium n-dodecylsulfate, $C_{12}H_{25}SO_4Na$, is a convenient example. The appropriate form of the Gibbs equation for this molecule in water is

$$\mathrm{d}\gamma = -2RT\Gamma_2^{(1)} \ln a_{\pm}, \tag{12.20}$$

where $a_{\pm}$ is the mean ion activity of the surfactant. Values of $\Gamma_2^{(1)}$ obtained in this way give the area $\sigma = 1/\Gamma_2^{(1)}$ occupied by a surfactant molecule at, say, the surface of water, or at the surface of an aqueous solution containing an inorganic salt (the factor 2 in equation 12.20 is eliminated in these circumstances). Values of σ obtained in this manner are compatible with a model in which surfactant anions are close-packed perpendicular to the aqueous surface: typical values are 53 Å^2 for $C_{12}H_{25}SO_4Na$ and 59 Å^2 for $C_7H_{15}CH(SO_3Na)COOC_6H_{13}$ in 0.01 m sodium chloride.

Adsorption at Liquid–Solid Interfaces

It will be recognized that binary mixtures with miscibility in all proportions have a symmetry not present when one component is a solute of limited solubility. Polymer adsorption from solution must take account of solvent qualities, e.g., good or theta, as already discussed in Chapter 11; further discussion is deferred to Chapter 13. The amount of a component adsorbed from solution onto a solid can be obtained by measuring the change in solution composition following immersion of the solid (often powdered) into the liquid. Solids are first of all cleaned by removal of adsorbed species by heating them under vacuum, they are then sealed in a container (ampoule) which is broken under the liquid surface. Adsorption processes are accompanied by enthalpy changes, in this case called the enthalpy of immersion of the solid, which can be interpreted in terms of liquid–solid interactions.

For a binary mixture of molecules 1 and 2 present in total amount n, equilibrated with mass m of solid adsorbent, the specific surface excess amount, $n(x_2^0 - x_2)/m$, where x_2^0 and x_2 are, respectively, the initial and final mole fractions of component 2 in the bulk mixture, can be measured. Figure 12.3 shows two typical forms of the corresponding surface excess isotherm. In this manner of representing adsorption, there is no adsorption of component 2 when $x_2 = 0$ and when $x_2 = 1$. In Figure 12.3(b) there is also a surface azeotrope signified by the zero adsorption at intermediate x_2. A monomolecular model of competitive adsorption on the solid surface can explain many systems, especially when the molecules occupy similar areas when adsorbed, e.g. benzene + cyclohexane. Unlike the case of adsorption

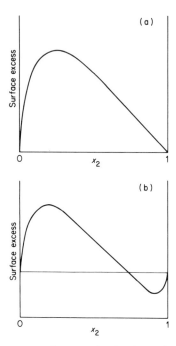

Figure 12.3. Two common types of composite isotherm for adsorption from solution on a solid. Type (a) is, for example, shown by benzene (1) + methyl acetate (2) on boehmite (alumina); and type (b) is shown by ethanol (1) + benzene (2) on carbon, with the surface azeotrope of bulk composition at x_2 $ca.$ 0.8

of a single component gas or vapour onto a solid, there is now a balance between the state when a molecule is attracted to the surface and surrounded on the other sides by molecules of kind 1 and 2, and the state where a particular molecule is surrounded by 1 and 2 in the bulk.[†] Many attempts have been made to deduce individual (component) isotherms from the corresponding composite or excess values using extra-thermodynamic assumptions and vapour-adsorption isotherms: they are outside the scope of this chapter. Many adsorption-from-solution isotherms, e.g. of amino acids in water, only cover a portion of the composition range before reaching solute saturation.

12.3 CAPILLARITY

There are many surface phenomena which do not depend on adsorption, but which are associated with the properties of fluid–fluid interfaces. They are related to the laws of capillarity.

---

†D. H. Everett, Spec. Period. Report, Chem. Soc., *Colloid Science*, **1**, 49 (1973).

When an interface between phases i and j is increased in area by $d\mathscr{A}^{ij}$ the work done on the system is

$$dW = \gamma^{ij}\,d\mathscr{A}^{ij} + \mathscr{A}^{ij}\,d\gamma^{ij}. \tag{12.21}$$

It will subsequently be assumed that γ^{ij} is constant. For finite reversible changes to a system involving liquid–solid, liquid–vapour and solid–vapour interfaces,

$$\Delta W = \gamma^{ls}\Delta\mathscr{A}^{ls} + \gamma^{lv}\Delta\mathscr{A}^{lv} + \gamma^{sv}\Delta\mathscr{A}^{sv}. \tag{12.22}$$

When no other changes are significant, e.g., there is no chemical reaction, and neglecting gravitational effects, the Helmholtz free energy change for a finite reversible change to a closed system is $\Delta F = \Delta W$.

Figure 12.4 shows a sessile drop of liquid on a solid surface. The contact angle θ is defined as the angle between the liquid–vapour interface and the solid, and is always taken, by convention, through the liquid, as shown. Contact angles generally pertain to the chemical composition of the three phases involved, but the contact angle is often found to vary over a range of values, due to contact-angle hysteresis. The limits of the range, $\theta_a > \theta > \theta_r$, as Figure 12.4 depicts, are the advancing angle θ_a and the receding angle θ_r depending on whether the static configuration has arisen from forward or backward movement, respectively, of the liquid phase over the surface. Possible causes of contact-angle hysteresis are chemical heterogeneity of the solid surface, surface roughness and the presence of impurities. To simplify the following discussion a unique contact angle θ will be assumed. The limiting case is $\theta = 0°$ when the liquid spreads over the solid surface. The

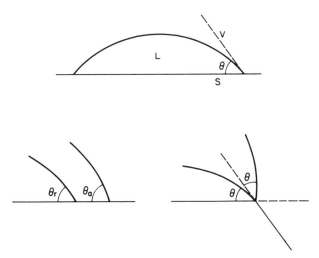

Figure 12.4. Sessile drop of liquid showing definition of contact angle θ. Contact-angle hysteresis can lead to two limiting values θ_a and θ_r. When the three-phase confluence is at a solid edge, 'hinging' is possible between limits where θ (assumed single value) is shown applying to the two solid surfaces meeting at the edge

cases of $\theta > 0°$ are divided, by convention, into wetting and non-wetting conditions:

$$0° < \theta < 90° \quad 90° < \theta < 180°$$
$$\text{wetting} \quad \text{non-wetting}$$

By an analysis involving the free energy change of the system, when the three-phase confluence is displaced with concomitant area changes, one of the general conditions for equilibrium when θ is not zero is obtained,

$$\gamma^{sv} = \gamma^{sl} + \gamma^{lv} \cos \theta, \tag{12.23}$$

which is known as Young's equation. A fluid–fluid interface may meet a solid surface at an edge which allows the interface some amount of 'hinging', without forming the contact angle θ (see Figure 12.4), and equation 12.23 is not applicable. The range of movement of the interface at a fixed edge will, however, be governed by θ with respect to the solid surfaces meeting at the edge, as Figure 12.4 also indicates.

Water, for example, has a contact angle of about 112° on solid paraffin and polyethylene surfaces, and would be said to be non-wetting; such surfaces are termed low-energy. On many clean inorganic solids, minerals and oxides and metals, known as high-energy surfaces, water spreads ($\theta = 0°$). A useful empirical concept, the critical surface tension of a solid surface, γ_c, has been found to be capable of discriminating between solids on which a liquid will spread from those on which a finite contact angle is formed. The quantity γ_c for a solid can be obtained by extrapolating plots of $\cos \theta$ against γ^{lv} for a series of liquids, e.g., alkanes, to $\cos \theta = 1$, or $\theta = 0°$. Other liquids with $\gamma^{lv} < \gamma_c$ will spread over the solid. Table 12.2 gives a few values of γ_c. Values of γ_c for a solid can be increased to give better wetting by chemical modification, e.g. by subjecting it to an electrical discharge (plasma) or by flaming. These are important considerations in achieving good adhesive joints; in general good wetting favours good adhesion. The semiqualitative classification into low- and high-energy surfaces arises because spreading will be spontaneous only when there is a free energy decrease, which means that the free energy associated with the creation of liquid–solid and liquid–vapour interfacial areas must be more than compensated by the decrease in solid–vapour interfacial free energy: equations 12.22 and 12.23.

Table 12.2. Critical surface tensions

Solid surface material	$\gamma_c/\text{mN m}^{-1}$
Polyethylene	31
Poly(vinyl chloride)	39
Polytetrafluoroethylene	18
Perfluorolauric acid monolayer on solid ($-CF_3$ exposed)	6

The interfacial tension between two fluid phases, where the interface behaves like a membrane in tension, has important consequences when the interface is curved. The mean surface curvature, which can vary over the interface, is

$$J = \frac{1}{r_1} + \frac{1}{r_2}, \tag{12.24}$$

where r_1 and r_2 are the principal radii of curvature of the surface. Figure 12.5 shows two simple interface shapes for which J can easily be determined, but for complicated interfacial configurations the evaluation of J is often difficult. At any position on a curved interface there is a component of γ acting inwards, giving rise to a pressure difference across the interface called the capillary pressure,

$$\Delta P = P^\alpha - P^\beta = \gamma J. \tag{12.25}$$

Equation 12.25 is usually referred to as Laplace's equation, although the underlying principle is also attributable to Young . The greater pressure is always on what can loosely be called the concave side of the interface. Liquid will rise up a capillary until the pressure difference given by 12.25 is balanced by the hydrostatic pressure of the column.

For a small spherical drop of liquid of radius r (see Figure 12.5) surrounded by air, the capillary pressure, given by equation 12.25, is $2\gamma/r$.

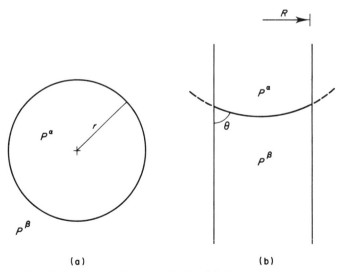

(a)   (b)

Figure 12.5. For the spherical drop of liquid (a) the mean surface curvature is $J = 2/r$. For liquid (b) in a solid cylinder (e.g., a pore) of radius R, where the contact angle is θ, the mean surface curvature is $J = -(2/R)\cos\theta$ (assuming a hemispherical surface). In both cases $P^\alpha > P^\beta$. The vapour pressure, equation 12.28, when one phase is a vapour or a mixture of vapour and gas, may need to be distinguished from these hydrostatic pressures

274

For a spherical soap bubble of radius r, the pressure difference is twice this value because there is a contribution from both the inner and outer surfaces of the soap film.

The force exerted on two solid particles separated by a fluid bridge, shown for two cylindrical solids in Figure 12.6, provides a more subtle problem, even neglecting gravity. The component of the fluid–fluid interfacial tension resolved axially, and acting around the edge of the solids, combined with the capillary pressure acting across the end faces of the solids, gives

$$f = 2\pi r \gamma \cos \phi - \pi r^2 \gamma J. \tag{12.26}$$

The interaction between soil particles often occurs via water bridges which can draw the particles together or hold them apart, depending on the volume and geometrical shape of the bridge. In a gravitational field the hydrostatic pressure in a fluid phase α decreases in the upward vertical direction z by $\rho^\alpha g z$ (Pascal's laws), and so the capillary pressure, or J at constant γ according to equation 12.25, must also vary. The most straightforward cases to deal with are those having an axis of rotation aligned in the vertical direction.† Figure 12.7 shows the meridian curves of some of the more common examples. In dealing with such static systems the quantities determining interface shape and stability are γ and the density difference $\Delta\rho$ between the fluid phases, which can be combined to define the capillary constant

Figure 12.6. Fluid bridge between the ends of two solid cylinders. Many types of fluid bridge are possible depending on the geometrical arrangement of the solids and the nature of the fluid phase. The force given by equation 12.26 is that acting on the upper solid where it joins the fluid of the bridge

---

†E. A. Boucher, *Reports Progr. Phys.*, **43**, 497 (1980).

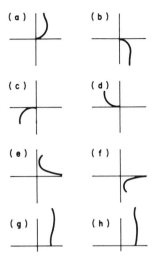

Figure 12.7. Eight types of meridian curve which on rotation about the vertical axis give the following axisymmetric fluid bodies in a gravitational field: (a) pendent drop, (b) emergent bubble, (c) sessile drop, (d) captive bubble, (e) raised holm, (f) submerged holm, (g) light bridge, and (h) heavy bridge

$$a = (2\gamma/\Delta\rho g)^{1/2}. \qquad (12.27)$$

An alternative definition omits the factor 2. What this effectively means is that a distance z in an actual system can be scaled by a to give a reduced or dimensionless quantity $Z = z/a$, regardless of the chemical composition. Furthermore, conditions of zero gravity, $g = 0$, can also be met by choosing two fluids of the same density, $\Delta\rho = 0$: the above dimensionless form is then not appropriate.

Hydrostatic pressures P^α and P^β must be distinguished from the vapour pressure p when one phase is a vapour in a gas. A plane liquid surface at temperature T will have above it a vapour pressure $p°$ when the hydrostatic pressure in the liquid phase is one atmosphere. For a liquid component l separated from the vapour phase by a curved interface, the hydrostatic pressure in the liquid will change, and so will its chemical potential μ_1. The change in chemical potential of the vapour phase component 1 compensating for this will be due to a change in fugacity, or, assuming ideal vapour, to a change from $p°$ to p given at equilibrium by

$$RT \ln p/p° = \gamma V_{\mathrm{m}} J, \qquad (12.28)$$

where V_{m} is the molar volume of the liquid. Equation 12.28 is known as Kelvin's equation. For a spherical liquid droplet of radius r surrounded by its vapour, $J = 2/r$: for a water drop at room temperature, $p/p° \approx 1.1$ for $r = 10$ nm.

A droplet of liquid will be unstable with respect to a plane liquid surface at the same temperature. It would evaporate, transferring by vapour diffusion to the plane surface. The sign of J in equation 12.28 determines whether $p > p°$ or $p < p°$. Examples of $p < p°$ are given by water in a glass capillary tube and water adsorbed (capillary condensation) by many porous solids, since the water–vapour interface is concave on the vapour side ($\theta < 90°$). On a plane solid surface in contact with water vapour there would be adsorbate increasing in thickness to give a layer a few molecules thick with increase in p, and eventually becoming thicker so that, at saturation, bulk condensation occurs ($\theta = 0°$ for simplicity).

Liquid Lenses and Spreading Phenomena

When a drop of liquid is placed on a second with which it is not miscible, three basic phenomena represented schematically in Figure 12.8 can occur: (a) the drop can form a lens; (b) the drop can spread to give a duplex film whose thickness depends on the area available for spreading; and (c) some of the drop can spread to form a monolayer on top of the liquid. A spreading liquid solvent can be used to facilitate the spreading of an insoluble substance, yielding an insoluble monolayer when the solvent evaporates.

The condition for mechanical equilibrium of the stable drop (Figure 12.8) is usually expressed as Neumann's triangle of forces,

$$\gamma^{wa} \cos \phi_1 = \gamma^{oa} \cos \phi_2 + \gamma^{ow} \cos \phi_3. \tag{12.29}$$

The shape and position of the lens will depend on the densities of the three fluid phases, and on the three interfacial tensions.

Spreading can be dealt with by introducing the spreading coefficient or tension S,

$$S = \gamma^{wa} - \gamma^{oa} - \gamma^{ow}. \tag{12.30}$$

If the full horizontal force γ^{wa} exceeds the sum of the other two, then spreading will occur:

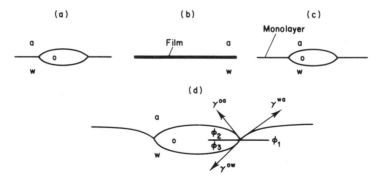

Figure 12.8. Representation of lens formation and spreading phenomena: (a) simple lens of one liquid on another; (b) spread film of liquid; (c) lens and spread monolayer; and (d) specification of a lens leading to equation 12.29

$$S > 0 \text{ spreading;} \quad S < 0 \text{ lens formation.} \qquad (12.31)$$

It is more satisfactory to consider the possibility of spreading in terms of the free energy change arising from area changes according to equation 12.22, but the conclusions will be the same for spreading conditions. Table 12.3 illustrates the use of S, and it shows how the situation can be altered when interfacial tensions change. The behaviour of benzene/water mixtures is unusual. Benzene after initially spreading retracts to a lens, leaving a monolayer of benzene on the remaining surface. A lower value of γ^{wa} due to an impure water surface would, other things being equal, reduce the chances of spreading occurring.

Spread monolayers of molecules such as hexadecanol can reduce the rate of evaporation from lakes. Artificially spread monolayers (as distinct from natural ones) may sufficiently affect lake surfaces to alter the ecology, e.g. by preventing mosquito larvae from maintaining a location for breeding.

Insoluble Monolayers

The aspect of insoluble monolayers to which this account is limited concerns the manner in which the two-dimensional pressure exerted by the monolayer depends on the surface area occupied by a known amount of substance. The use of a Langmuir–Blodgett or Langmuir–Adam trough for this purpose has a considerable history. It is still used for understanding intermolecular interactions, particularly (these days) those of biologically significant species. Figure 12.9 shows a highly schematic representation of the principle of operation.

The insoluble monolayer is to be contained between the movable barrier and the float attached to the torsion balance. The substance to be spread is dissolved in e.g., petroleum ether, and placed in small quantities at various locations on the clean water surface. A uniform layer of adsorbate should be obtained when the solvent has evaporated. The presence of the adsorbed layer reduces the surface tension relative to that, γ_0, of the still clean water on the other side of the float, giving a spreading pressure π,

Table 12.3. Behaviour of oily liquids on water†

Liquid	$\gamma^{oa}/\text{mN m}^{-1}$	$\gamma^{ow}/\text{mN m}^{-1}$	$S/\text{mN m}^{-1}$	Comment
n-Octane	21.8	50.8	+0.2	Can just spread
n-Octanol	27.5	8.5	+36.8	Spreads
n-Hexadecane	30.0	52.1	−9.3	Lens, no spreading
Benzene	28.9	35.0	+8.9 to −1.4	Initially spreads, then gives lens with surrounding monolayer

†Using $\gamma^{wa}/\text{mN m}^{-1} = 72.8$ at 293 K.

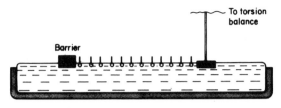

Figure 12.9. Schematic representations of the principle of operation of the Langmuir trough

$$\pi = \gamma_0 - \gamma. \tag{12.32}$$

The force needed (measured with the torsion mechanism) to maintain the float in position as the movable barrier slides to compress the adsorbed layer is (per unit length of float) a direct measure of π. The area available to the amount of spread substance is given from the geometrical area of the enclosed surface. Plots of π against the area per molecule can be constructed, as shown schematically for stearic acid and myristic acid (octa- and tetra-decanoic) in Figure 12.10. Plots are usually interpreted in terms of

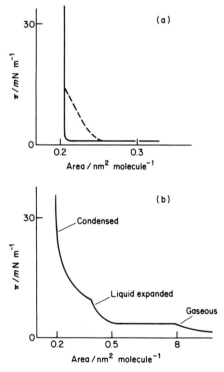

Figure 12.10. Dependence of surface pressure, π, on the area available per molecule of spread substance, shown schematically for (a) stearic acid on water (—) and dilute HCl (– –) at 293 K, and (b) myristic acid on $0.1 \, mol \, dm^{-3}$ hydrochloric acid at 288 K. The area scale in (b) is distorted to show the main features—see text

an analogy to a hypothetical two-dimensional gas which on compression changes 'state'. Straight chain fatty acids, e.g. stearic, at room temperature give very low pressures when there is a large area available per molecule: the oriented molecules form two-dimensional clumps or islands due to hydrocarbon chain interactions. On compression they pack tightly and the surface pressure rises steeply. Myristic acid molecules, by comparison, separate at large available areas and behave like a two-dimensional gas. This can be compressed to give a so-called liquid-expanded region, and further compression eventually produces a condensed state: the film cannot withstand further compression and would buckle. The phospholipid molecules discussed in Chapter 13 have been studied in this manner.†

12.4 NUCLEATION PHENOMENA

Nucleation initiates the onset of a phase change, e.g. vapour → liquid or liquid → solid; the discussion of the phase diagrams in Chapters 1, 4 and 7 assumed that there are no impediments to the changes. In practice, different phase changes can cover a wide range of time-scale due to energy barriers, and the time-scale of a particular change can be varied by altering the height of the energy barrier. While microscopic treatments of nucleation are available‡, it is instructive to deal here with its macroscopic aspects.

A common illustrative experiment involves observing crystal growth from the melt of sodium thiosulfate, which once started occurs rapidly from a few locations. These locations are the positions of nuclei which, as will be seen, depend on the cleanliness of the system; crystal growth would be greatly helped by adding a few crystals to the supercooled melt (seeding). The slow crystallization of clear honey represents a situation where nucleation is not readily achieved. Many natural phenomena, e.g. cloud (liquid droplet) formation and ice (snow or hail) formation, involve nucleation phenomena. The production of a new phase such as a liquid from vapour, or solid from liquid, even though thermodynamically favourable in terms of the chemical potentials of the bulk phases, requires a mechanism often entailing an embryo or nucleus of the new phase. When only a single component or a phase of uniform composition is involved the process is termed homogeneous nucleation. When the embryo of the new phase forms on a different (inert) solid surface, the process is termed heterogeneous nucleation.

Figure 12.11 depicts both kinds of process. Clearly all the molecules of a liquid cannot suddenly adopt positions required to form the solid. By forming tiny clusters an interface is produced between solid and liquid with

---

†M. N. Jones, *Biological Interfaces*, Elsevier, London and Amsterdam (1975).
‡A. C. Zettlemoyer (ed.), *Nucleation*, Dekker, New York (1969).

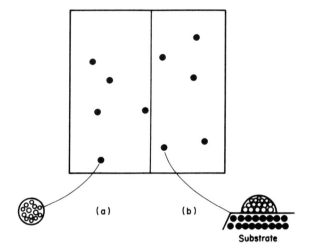

Figure 12.11. Schematic representation of (a) homogeneous nucleation, and (b) heterogeneous nucleation on a solid substrate

an increase in free energy. For liquid droplet formation from a vapour, for instance,

$$\Delta G = \Delta G_{\text{bulk}} + 4\pi r^2 \gamma^{\text{lv}}. \tag{12.33}$$

Figure 12.12 shows how ΔG depends on the radius r of the cluster for several values of the supersaturation ratio S of the vapour. There is a critical nucleus size r^* corresponding to the maximum ΔG^* in the free energy change. Only if a cluster of atoms or molecules can grow and slightly exceed this critical size will the subsequent spontaneous growth of the new phase be possible. The rate j of nucleation depends strongly on ΔG^*,

$$j \sim \exp(-\Delta G^*/RT), \tag{12.34}$$

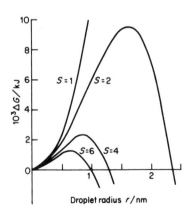

Figure 12.12. Dependence of ΔG for water droplet formation on droplet size for various saturation ratios S

and any way of reducing ΔG^* will increase the rate of growth to a size $r > r^*$. In heterogeneous nucleation the corresponding energy barrier is lowered by energetically favourable cluster formation on the solid surface. The dependence on the exponent in equation 12.34 is often so sharp that a small temperature change can result in very rapid nucleation and growth when effectively the rate was previously zero. Solid atmospheric particles (for example) nucleate ice from supercooled water vapour in the presence of liquid droplets at a characteristic threshold temperature: for the artificial cloud-seeding agent silver iodide, the threshold temperature is 267 K.

Chapter 13

Colloidal Systems

13.1 INTRODUCTION

A molecule in a surface layer is in a different environment from otherwise identical molecules in the bulk material because it is not subject to the same field of intermolecular forces. The enthalpy, entropy and free energy of surface layers will generally be different from those of the bulk material, and this must be taken into account in any thermodynamic treatment of a system in which the surface is significant. When at least one dimension of a phase is sufficiently small, the surface may constitute a substantial fraction of the total material, and the surface properties may dominate behaviour; materials in this condition are usually (but not always) part of two-phase systems, and are known as colloids. It is impossible to impose a sharp borderline of size separating colloidal and conventional systems but, as a rough guide, one can take the upper limit of the critical dimension of a colloidal material to be about 10^{-6} m or $1\,\mu$m. In most cases, a colloid has a highly subdivided substance (or disperse phase) widely distributed as a solid or liquid throughout a liquid or gas, the continuous phase, but there are other types, for example, very thin membranes where there is no disperse phase. Many colloids occur in Nature, both in animate and inanimate environments, and many others are of considerable commercial or domestic importance.

Apart from stipulating that there must be at least one dimension in the micron region, it is impossible to generalize about the nature of colloidal systems. Some are true solutions (lyophilic colloids), such as polymers in good solvents; others are dispersions of insoluble material (lyophobic colloids), such as silver iodide in water. Yet others are emulsions, foams, and even membranes; Table 13.1 lists most of the commonly-encountered types of colloidal system. They all have at least one phase with a very large surface/volume ratio, and both their physical and chemical properties differ from those of the same substances in non-colloidal form.

Table 13.1. Some types of colloids

Phase		Name	Example
Continuous	Disperse		
Gas	Liquid	Liquid aerosol	Mist, spray
Gas	Solid	Solid aerosol	Dust, fog
Liquid	Gas	Foam	Froth
Liquid	Liquid	Emulsion	Milk, pharmaceutical bases
Liquid	Solid	Colloidal suspension (sol)	Gold sol
Solid	Gas	Solid foam	Polyurethane foam
Solid	Liquid	Solid emulsion	
Solid	Solid	Solid suspension	TiO_2 in plastics

13.2 FORMATION OF COLLOIDAL DISPERSIONS

Most of the methods of forming colloids are based on either the aggregation of more highly subdivided material or the comminution (reduction to very small particles) of bulk samples. Breaking down by grinding tends to produce particles at the upper end of the colloid size range and with a broad size-distribution. Some chemical reactions, carried out under carefully controlled conditions, yield a colloidal product rather than a bulk precipitate: thus, reaction of silver nitrate with potassium iodide can be made to furnish a stable colloidal solution (or sol) of silver iodide, and $HAuCl_4$ can be reduced by formaldehyde to form a colloidal solution of elemental gold. Sulfur sols can be prepared by grinding solid sulfur with a liquid, by adding a dilute solution of the element in ethyl alcohol to water, or by the reduction of sodium thiosulfate.

The nucleation step is often important in determining whether or not a colloidal solution is formed in a precipitation procedure; if nucleation is rapid, it is sometimes possible to form a colloid with a narrow size-distribution in the particles of the disperse phase. The nucleation of a gold sol is best achieved by reducing a solution of $HAuCl_4$ (neutralized with potassium carbonate) with white phosphorus, while polystyrene latices with narrow particle size-distributions (so-called uniform or monodisperse systems) can be prepared by carefully-controlled emulsion polymerization, a process described in textbooks of polymer chemistry.

13.3 COLLOID STABILITY

Some colloids, such as solutions of polymers or aggregates of surfactant molecules, may be thermodynamically stable but many are not, even though they may remain apparently unchanged for very long periods of time. One might expect collisions between particles of the disperse phase (induced by

Brownian motion) to lead to flocculation or coagulation (see note regarding the meanings of these terms on page 206) but this is not necessarily the case because aggregation may be prevented by repulsive forces. There are two recognised mechanisms for conferring stability on the disperse phase; these are (i) the formation of an electrical double-layer at the particle surface and (ii) the adsorption of neutral material at the surface, which leads to steric stabilization.

The Electrical Double-layer

Proteins and similar molecules dispersed in water can acquire an electric charge, dependent on the pH of the medium, by self-ionization to $-NH_3^+$ and/or $-COO^-$ groups. Electrostatic interaction will ensure that whichever ionic species is present will have counterions of the opposite charge immediately around it. Since the charge on the dispersed particles is attached to a surface, the counterions tend to form a parallel second layer, and the combined system is known as an electrical double-layer or Stern layer. Particles dispersed in water can also acquire charge by absorbing foreign ions from the surrounding solution, and a double-layer of charge similarly forms close to the particle surface. As the electrostatic forces act over a long distance, counter-ions are not limited to the double-layer but their concentration will decrease with increasing distance from the surface. The distribution of ions beyond the initial double-layer is known as the diffuse layer; it is more likely to be disturbed by thermal motion and is analogous to the ionic atmosphere around a central ion in the Debye–Hückel concept (page 206 and following pages).

The type of ion at the particle surface is called the potential-determining ion, and the form of the variation of the electric potential ψ with distance x from the surface dominates the behaviour of the colloid. Naturally-occurring aqueous colloidal dispersions are usually, but not always, negatively charged, probably because large polarizable anions are more strongly adsorbed than (hydrated) cations. Models for the electrical double-layer treat the Stern and diffuse layers separately. The simplest treatments, e.g. that proposed by Gouy and Chapman, assume that the potential-determining ions are on a plane surface, and that the remainder of the ions are point charges in a Boltzmann distribution, as represented schematically in Figure 13.1. In the diffuse layer, the net density of charge is related to ψ by Poisson's equation (see page 219) together with the approximation that $Ze\psi_0/2kT \ll 1$, where $\psi_0 = \psi$ at $x = 0$. (Here, Z is the number of unit charges on an ion and e is the unit electrical charge.) Thus:

$$\psi = \psi_0 \exp(-k\psi), \qquad (13.1)$$

where $1/\kappa$ is the thickness of the diffuse layer, given by

$$\kappa = \left(\frac{2e^2 n_0 Z^2}{\varepsilon_0 \varepsilon_r kT}\right)^{1/2}. \qquad (13.2)$$

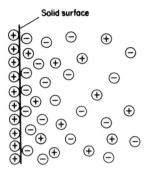

Figure 13.1. Schematic representation of the ionic distribution adjacent to a positively charged surface in the Gouy–Chapman model

In this equation, n_0 is the number density of ions (the product of the molar electrolyte concentration and Avogadro's number) and ε_r is the relative permittivity of the medium. The variation of electric potential with distance from the surface implied by this model is shown in Figure 13.2(a).

Close to the solid surface, the finite size of the ions and their level of hydration vitiate the assumption of point charges. The centre of a hydrated ion cannot approach the charged surface more closely than its radius; this defines the inner boundary of the Stern layer, as shown in Figure 13.3. The presence of these ions modifies the dependence of ψ on x, as shown in Figure 13.2(b) but, further from the surface (in the diffuse layer), the exponential function is a satisfactory description of the decay of ψ from ψ_δ (but not from ψ_0). The specific adsorption of ions can complicate the situation, even to the point of reversing the sign of the charge at the Stern layer. Apart from their effect on $\psi(x)$, the specific adsorption of hydrated

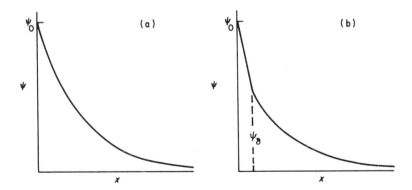

Figure 13.2. Dependence of the electrical potential ψ near the colloid particle on distance x: (a) for the diffuse region, and (b) for the Stern model of Figure 13.3 with the diffuse model applying except very close to the surface (the precise dependence of ψ on x near the surface depends on the system, but that shown is thought to be the usual behaviour)

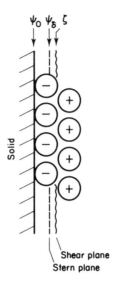

ψ_0 ψ_δ ζ

Solid

Shear plane
Stern plane

Figure 13.3. Schematic representation of the region very close to the solid, showing the Stern plane where the potential is ψ_δ and the shear plane where it is ζ

ions affects the mobility of the solution with respect to the surface; such movement takes place at the shear plane shown in Figure 13.3, where the potential is labelled ζ. This zeta potential is the most important measurable characteristic of a charged surface in a colloidal system.

The electrical double-layer stabilizes colloids against flocculation by establishing a potential-energy barrier; this arises from the overlap of regions of $\psi(x)$, as is shown schematically in Figure 13.4. It is thought that, under some circumstances, a shallow secondary minimum can exist further out than the maximum in Figure 13.4, and give rise to loose readily-reversible flocculation. (Some authorities distinguish between coagulation and flocculation, according to whether the change takes place at the primary or secondary minimum but this distinction is not general.) The level of protection afforded by the maximum in the potential-energy barrier is strongly dependent on its height; moreover, addition of unreactive electrolyte to the aqueous phase increases the concentration of ions in the vicinity of the charged surface, causing ψ to decrease more rapidly with x than would otherwise be the case, i.e. it compresses the double-layer, leading to coagulation. Increase in ionic charge, using multiply-charged ions, has a dramatic effect of a similar kind, a result embodied qualitatively in the Schulze–Hardy Rule, which states that flocculation is most efficiently brought about by the addition of ions of opposite charge from the surface of the colloid particles, especially if those ions are highly charged. The critical concentrations of 1:1, 2:1 and 3:1 electrolytes for inducing the coagulation of a negatively-charged silver iodide sol are, respectively, 0.140, 2.5×10^{-3} and

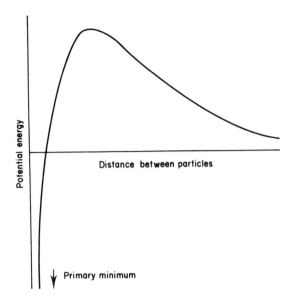

Figure 13.4. Potential energy barrier for two stabilized colloid particles. The primary minimum and 'hard-sphere' repulsion curve are omitted

7.0×10^{-5} mol l^{-1}, i.e. a lanthanum salt is more than 10^3 more effective than the corresponding potassium salt.

The modern theory of the stability of lyophobic sols has its foundations in the work of Derjaguin and Landau and of Verwey and Overbeek. They made theoretical investigations of the interactions between model systems, e.g. two charged spheres or two charged parallel plates, and extended the results to colloids. The substance of these studies are often known as the DLVO theory.

The electrical double-layer is responsible for the movement of particles of the disperse phase material when they are subjected to an electric potential gradient and, conversely, an electric potential gradient is created when charged colloidal particles are forced to move relative to a surface. The types of behaviour associated with these effects are known as electrokinetic phenomena, and there are four such kinds of experiment: electrophoresis; electro-osmosis; streaming potential; and sedimentation potential. Some brief notes are appended but the interested reader is referred to specialist textbooks for details.

Electrophoresis occurs when a potential gradient is applied to a colloidal system, and the equilibrium velocity of movement of the particles is measured, as shown in Figure 13.5. The ζ-potential can be estimated from such measurements, and charged species can be separated from one another and from uncharged material in this way. Parallel techniques employing a gel or paper as the vehicle for movement are also often used, especially for

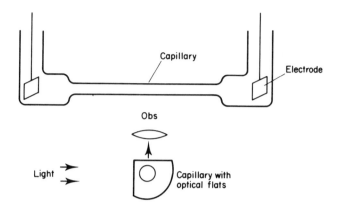

Figure 13.5. Schematic arrangement for particle electrophoresis. The central portion of the capillary joining the electrode compartments is shown in enlarged section. Scattered light from the particles is viewed at right angles to the incident beam through a microscope

biochemical applications. Electrophoretic mobility can be used to assess the effect of modification of the particle surface, e.g. by the adsorption of ions or surfactants, or the effect of pH. An important characteristic of colloids is the value of the pH at which the particles carry zero net charge, i.e. where the ζ-potential becomes zero.

Electro-osmosis is the phenomenon of flow which occurs when a potential difference is applied to a liquid within a solid plug formed by the compaction of solid particles bearing a charged surface; an electro-osmotic pressure develops. This phenomenon interferes with electrophoresis because the walls of the glass in contact with the solution will bear a charge; this causes unidirectional flow to occur near the walls, compensated by return flow along the tube axis. Calculation shows that the two effects cancel at a fraction 0.146 of the bore diameter in from the wall, and the true electrophoretic result can only be obtained from observations on the movement of particles at precisely this point.

Streaming potential is the reverse of the electro-osmotic effect, i.e. it is the development of an electric potential when a liquid is forced through a plug or porous bed.

Sedimentation potential is the phenomenon of the development of an electric potential difference between the top and bottom of a vessel in which charged particles are settling.

Steric Stabilization

Colloids may be stabilized against coagulation by the adsorption from the solution of bulky uncharged molecules, usually polymers or non-ionic surfactants. Block copolymers (see Figure 11.1) are particularly effective in this respect if one block has an affinity for the colloid particle while another

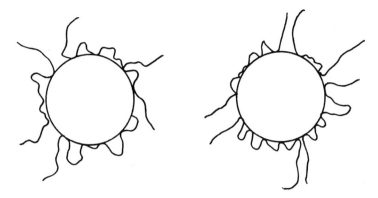

Figure 13.6. Schematic representation of polymer adsorption on solid particles. Some segments of each polymer molecule adsorb but others are surrounded by solvent

block interacts favourably with the liquid medium. Figure 13.6 shows schematically how polymer molecules can cover the surface of a colloidal particle and, at the same time, penetrate significantly into the liquid phase. If individual polymer molecules tend to adopt the random-coil configuration, interpenetration following contact between two colloid particles would lead to some mutual alignment with consequent loss of entropy, and this affects the free energy of the system in a way which disfavours coagulation sufficiently to provide a barrier which effectively stabilizes the colloid.

Adsorption of polymer molecules on colloids can also have the opposite effect by the process called bridging flocculation, where different segments of a single macromolecule adsorb simultaneously on to two or more colloid particles, leading to an attractive force which results in aggregation.

13.4 ASSOCIATION COLLOIDS

In some circumstances, certain types of molecules placed in a liquid medium spontaneously form aggregates. This is particularly the case when the liquid is water and the molecules in question are of the amphipathic type, i.e. they are surfactants. The aggregates formed by water-soluble surfactants, such as sodium dodecyl sulfate (SDS), are called micelles, and are illustrated in Figure 13.7. Micelles exist only when the concentration of the surfactant is in excess of a definite minimum value, known as the critical micelle concentration (c.m.c.); for SDS, the c.m.c. is approximately $0.08 \, \text{mol} \, l^{-1}$. Colloids formed as a consequence of this type of aggregation are called association colloids, in contrast to solutions of macromolecules, where the colloidal character reflects the properties of individual molecules rather than aggregates of molecules.

As was explained in Chapter 12, the Gibbs adsorption equation provides a

290

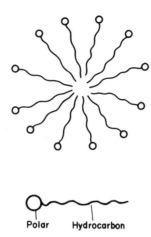

Polar   Hydrocarbon

Figure 13.7. Representation of a spherical micelle. The outer surface may consist of an electrical double layer or may be non-charged, depending on the surfactant species

basis for using the variation of surface tension with concentration to estimate the amount of surfactant adsorbed at the liquid–vapour interface. The surface tension increases with concentration up to the c.m.c. but thereafter remains constant because additional solute is being used to build micelles. Many other properties of the system, e.g. electrical conductivity and osmotic pressure, also show a discontinuity at the c.m.c..

The interiors of micelles in aqueous media are hydrophobic, hence they may act as traps for oily substances. This type of behaviour is called solubilizing; the use of surfactants and soaps in washing depends on the ability of micelles to solubilize the dirt removed from a surface. In the body, bile acids and salts solubilize fatty substances. The same property has uses in chemistry: reactive organic molecules can react more rapidly with one another when solubilized because they then exist at a much higher local concentration: also, the commercially-important process of emulsion polymerization depends on micelles to provide a locus for reaction.

Micelles of simple surfactants are believed to be roughly spherical in shape at concentrations just above the c.m.c.. For straight-chain surfactants with 12–14 carbon atoms, the diameter is about 5 nm, corresponding to a content of 30–50 molecules. The formation of a typical micelle can readily be pictured as consisting of two consecutive processes: (i) the assembly of hydrocarbon 'tails' and (ii) the liberation of water molecules bound to the surfactant molecule, enabling more structured water (i.e more extensively hydrogen-bonded water) to form. The forces of attraction between the hydrocarbon moieties are not significantly greater (and may be less) than those between an individual hydrocarbon component and the surrounding water but they are much less than the water–water attractions and the interaction of water with the hydrophilic component of the surfactant. Thus, the driving force for this process is the overall increase in polar interactions even though it is known as the hydrophobic effect, which often plays an important role in hydrocarbon–water systems.

Many surface-active molecules, for example those which form the insoluble monolayers described in Chapter 12, are insoluble in water, nevertheless they can be dispersed into aggregates in water by strong mechanical agitation or by the application of ultrasonic radiation. Such fatty aggregates are called liposomes, and similar biological entities are called vesicles. Figure 13.8 depicts the essential features of liposome structure, showing its bilayer nature and the various shapes, including concentric forms, that are known. The insoluble surfactants have a more complex structure than the simple (soluble) amphiphiles with a single hydrocarbon chain and a single hydrophilic head, indeed it is common for them to have at least two hydrocarbon portions. A naturally-occurring phospholipid, dispersed in at least an equal quantity of water, forms aggregates with concentric cylindrical bilayers which transform into spherical shape on gentle shaking.

Liposomes will form only at temperatures below the so-called chain-melting temperature of the phospholipid; solid phospholipids exhibit polymorphism, including mesomorphic phases, and the chain-melting temperature is the point at which the change from crystalline to liquid-crystalline phase takes place. Lyotropic mesophases also form when phospholipids are heated with water, and analogous transformations occur in monolayers spread on a water surface.

It is believed that biological membranes comprise lipoprotein layers containing protein molecules, with other proteins coating the inner and outer surfaces. Liposomes can thus be regarded as model membranes and, in this context, they have been used to study important biological membrane properties, such as the diffusion of ions from the interior to the external

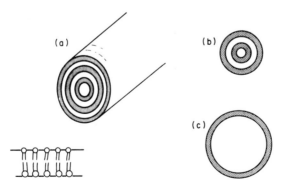

Figure 13.8. Bilayer structures as (a) cylindrical aggregates, (b) spherical concentric layers and (c) liposomes. The schematic bilayer arrangement shows the phospholipid molecules, e.g., phosphatidylcholine (lecithin)

aqueous region. They are also of interest pharmacologically as carriers of drugs. Their liquid-crystalline nature means that they are very sensitive to small changes in the body temperature, and they can react to local high temperature in the blood stream by breaking down to release a drug at the site of infection.

13.5 MACROMOLECULES AS COLLOIDS

The essential physical chemistry of polymer molecules is discussed in Chapter 11. The size of many individual macromolecules is such that they fall into the colloid size-range, so that solutions of polymers exhibit all the characteristic properties of colloids. For example, the dimensions of a typical randomly-coiled macromolecule are likely to be comparable to the wavelength of visible light, and we have already seen that light-scattering provides one of the best ways of measuring the molecular weight of a polymer. We have also noted that relatively small polymer molecules confer stability on other colloids when adsorbed on their surfaces. Biological macromolecules are often insoluble in water but may disperse in an aqueous medium.

13.6 AEROSOLS, FOAMS AND EMULSIONS

Some of the account given above in this chapter applies directly to the types of colloidal system known scientifically as liquid-in-gas, gas-in-liquid and liquid-in-liquid colloids but more colloquially as aerosols, foams and emulsions, respectively. In addition, they have some other distinctive properties, described below.

Aerosols can be formed by condensing vapour or by dispersing a liquid by forcing it at high velocity through a nozzle, as from an 'aerosol can'. The condensation of a vapour relies on nucleation, homogeneous or hetero-geneous; for example, the formation of clouds in air, when the water vapour pressure is close to the saturation level (the dew point), is due to atmospheric dust particles acting as nuclei for the vapour-to-liquid phase change.

Liquid droplets have an equilibrium vapour pressure, given by Kelvin's equation, 12.28, greater than the surrounding pressure, the difference being significant if the radius of the drop is less than about 1 μm.

The relative stability of aerosols depends on the efficiency of collisions in causing coalescence, and on the ease with which evaporation or further condensation can occur, transport mainly taking place by diffusion. Smog particles (atmospheric droplets of a different kind) acquire some resistance to evaporation by accumulating a surface coating of oily substance. Cloud droplets sometimes contain dissolved salts, which partially offset the elevation of the vapour pressure according to Kelvin's equation (see page 275); thus, they are often able to withstand supercooling to well below 0 °C without converting to ice.

Foams are broadly of two types: (i) those with a large gas-to-liquid ratio, as in a froth of polyhedral gas cells or in soap bubbles, and (ii) those with a low gas-to-liquid ratio, where the bubbles remain separate and spherical, as when a sealed bottle of a carbonated drink is opened.

The stability of the more concentrated foams can be understood by considering a single soap film drawn from the bulk liquid. Initially, drainage and thinning occur by gravity (like water running down a window pane) but, subsequently, liquid is squeezed out in all directions by capillary attraction, producing further thinning (see equation 12.25). The presence of surfactant molecules contributes to the viscosity of the liquid and, more importantly, it stabilizes the film, e.g. by formation of electric double-layers on opposite sides; in very thin films, e.g. black soap films about 5 nm thick, the hydrocarbon portions of the surfactant molecules are stacked end-to-end in a bilayer. When viewed with reflected white light, coloured interference fringing denotes film-thinning until constructive interference occurs at all visible frequencies.

The adsorption of surfactant molecules, such as proteins, by foams helps to stabilize the head on beer. It also provides a method of purification of surfactant solutions by foam fractionation. In this technique, impurities adsorbed at the liquid-surface interface can be skimmed off when gas is bubbled through the liquid to enhance foaming. The production of gas bubbles in a liquid also provides a basis for mineral flotation; some solid particles are carried away as a result of becoming incorporated into the liquid-bubble interface. (Such systems are not strictly foams because the system is neither static nor stable.)

Emulsions consist of droplets of one liquid dispersed in the continuous phase of another, one of the liquids usually being water. The two principal types are called oil-in-water (o/w) and water-in-oil (w/o) emulsions, the term oil being used in its most general sense to mean a liquid immiscible with water. Simple hydrocarbons do not form an emulsion when shaken with water unless a surface-active compound is also present; thus, a small amount of a soap will stabilize an emulsion of benzene and water. The o/w and w/o types of emulsion differ in their properties, many of which reflect the nature of the continuous phase: o/w emulsions are much better conductors of electricity; they tend to have a creamy consistency while the w/o variety are usually greasy to the touch. The latter are used extensively in pharmaceutical preparations, and many types of emulsions feature in food: ice-cream, mayonnaise and margarine are examples. Frequently, the particles in emulsions exceed the nominal 1 μm upper size-limit for colloids, and they are rarely transparent because intense light-scattering takes place; however, a variety called microemulsions constitute an exception to this statement.

The stability of emulsions is intimately associated with adsorption at the oil–water interface, involving either the electric or steric mechanism. Emulsions can be stabilized *inter alia* by proteins, glucosides and lipids, including gelatine, pectin and albumin. They can also be stabilized by the capillary attraction of small solid particles to the oil–water interface, their favoured position with contact angles greater than zero. These solid particles may themselves acquire an electrical double-layer, thus enhancing their ability to maintain droplets apart and reduce the likelihood of coalesence. Emulsions are systems of considerable scientific and practical interest but Art is usually required to reinforce Science in the attainment of satisfactory emulsion formulations.

Bibliography

This bibliography contains most textbooks on liquids and solutions and a selection of textbooks on some of the subjects that we have dealt with such as liquid crystals and colloids. Most of these books are more suitable for post-graduate than for undergraduate reading—even some with titles beginning 'Introduction to . . .'. They will, however, be useful for students who wish to look at certain topics in more detail than we have provided.

Adamson, A. W., *Physical Chemistry of Surfaces*, 3rd edn, Interscience, Chichester (1977)

Albert, A., and Serjeant, E. P., *The Determination of Ionization Constants: a Laboratory Manual*, Chapman and Hall, New York (1971)

Amis, E. S., and Hinton, J. F., *Solvent Effects on Chemical Phenomena*, Vols. 1 and 2, Academic Press, London and New York (1973)

Aveyard, R. and Haydon, D. A., *An Introduction to the Principles of Surface Chemistry*, C.U.P., Cambridge (1973)

Barker, J. A., *Lattice Theories of the Liquid State*, Pergamon Press, Oxford (1963)

Barton, A. F. M., *The Dynamic Liquid State*, Longman, London (1974)

Bell, R. P., *Acids and Bases: their Quantitative Behaviour*, 2nd edn, Methuen, London (1969)

Blackburn, T. R., *Equilibrium: a Chemistry of Solutions*, Holt, Rinehart and Winston, New York (1969)

Blander, M. (Ed.), *Molten Salt Chemistry*, J. Wiley, London and New York (1964)

Boublik, T., Nezbeda, I. and Hlavaty, K., *Statistical Thermodynamics of Simple Liquids and their Mixtures*, Elsevier, London and Amsterdam (1980)

Brown, G. H. (Ed.) *Liquid Crystals* (conference report), Kent State Univ., Ohio (1968)

Buckingham, A. D., Lippert, E., and Bratos, S., *Organic Liquids*, J. Wiley, Chichester (1978)

Burgess, J., *Metal Ions in Solution*, Ellis Horwood, Chichester (1978)

Butler, J. N., *Solubility and pH Calculations: the Mathematics of the Simplest Ionic Equilibria*, Addison-Wesley, New York (1964)

Butler, J. N., *Ionic Equilibrium, a Mathematical Approach*, Addison-Wesley, New York (1964)

Charlot, G., Collumeau, A., and Marchon, M. J. C., *Selected Constants: Oxidation—Reduction Potentials of Inorganic Substances in Aqueous Solution* (IUPAC) Butterworths, London (1971)

Coetzee, J. F., and Ritchie, C. D., *Solute—Solvent Interactions*, Marcel Dekker, New York (1969)

Cole, G. H. A., *An Introduction to the Statistical Theory of Classical Simple Dense Fluids*, Pergamon Press, Oxford (1967)

Conway, B. E., *Electrochemical Data*, Elsevier, London and Amsterdam (1952)

Conway, B. E., and Barradas, R. G., *Chemical Physics of Ionic Solutions* (symposium), J. Wiley, London and New York (1966)

Covington, A. K. and Jones, P., *Hydrogen Bonded Solvent Systems* (symposium), Taylor and Francis (1968)

Croxton, C. A., *Liquid State Physics—a Statistical Mechanical Introduction*, C.U.P., Cambridge (1974)

Croxton, C. A., *Introduction to Liquid State Physics*, J. Wiley, Chichester (1975)

Croxton, C. A., *Statistical Mechanics of the Liquid Surface*, J. Wiley, Chichester (1980)

Dack, M. R. J. (Ed.), *Solutions and Solubilities*, Parts I and II, J. Wiley, Chichester (1976)

Durrans, T. H., *Solvents*, 8th edn, Chapman and Hall, New York and London (1971)

Egelstaff, P. A., *An Introduction to the Liquid State*, Academic Press, New York and London (1967)

Eisenberg, D., and Kauzmann, W., *The Structure and Properties of Water*, Clarendon Press, Oxford (1969)

Everett, D. H., *Basic Principles of Colloid Science*, The Royal Society of Chemistry, London (1988).

Eyring, H., and Jhon, M. S., *Significant Liquid Structures*, J. Wiley, London and New York (1969)

Faber, T. E., *Introduction to the Theory of Liquid Metals*, Cambridge University Press, Cambridge (1972)

Falkenhagen, H., *Electrolytes*, Clarendon Press, Oxford (1934)

Faraday Society Discussion, **43**, Structure and properties of liquids (1967)

Faraday Discussions of the Chemical Society, **66**, Structure and motion in molecular liquids (1978)

Fisher, I. Z., *Statistical Theory of Liquids*, Univ. of Chicago Press, Chicago (1964)

Flory, P. J., *Principles of Polymer Chemistry*, Cornell, New York (1953)

Franks, F. (Ed.), *Physico-chemical Processes in Mixed Aqueous Solvents*, Heinemann, London (1967)

Franks, F. (Ed.), *Water: a Comprehensive Treatise*, vols. 1–7, Plenum Press, London and New York (1972)

Gennes, P. G. de, *The Physics of Liquid Crystals*, Clarendon Press, Oxford (1974)

Gerrard, W., *Solubilities of Gases and Liquids, a Graphic Approach*, Plenum Press, London and New York (1976)

Gerrard, W., *Gas Solubilities*, Pergamon Press, Oxford (1980)

Gold, V., *pH Measurements: their Theory and Practice*, Methuen, London (1956)

Gray, G. W., *Molecular Structure and the Properties of Liquid Crystals*, Academic Press, New York and London (1962)

Guggenheim, E. A., *Mixtures, the Theory of the Equilibrium Properties of Some Simple Classes of Mixtures, Solutions and Alloys*, Clarendon Press, Oxford (1952)

Guggenheim, E. A., and Stokes, R. H., *Equilibrium Properties of Aqueous Solutions of Single Strong Electrolytes*, Pergamon Press, Oxford (1969)

Hamer, W. J. (Ed.), *The Structure of Electrolytic Solutions* (symposium), J. Wiley, London and New York (1959)

Hansen, J. P., and McDonald, I. R., *Theory of Simple Liquids*, Academic Press, London (1976)

Harned, H. S., and Owen, B. B., *The Physical Chemistry of Electrolytic Solutions*, Reinhold, New York (1943 and 1958)

Hildebrand, J. H., Prausnitz, J. M. and Scott, R. L., *Regular and Related Solutions*. Van Nostrand-Reinhold, New York (1970)

Hildebrand, J. H., and Scott, R. L., *The Solubility of Non-electrolytes*, Dover, London (1964)

Hildebrand, J. H., and Scott, R. L., *Regular Solutions*, Prentice-Hall, Englewood Cliffs, New Jersey (1962)

Hirschfelder, J. O., Curtis, C. F., and Bird, R. B., *Molecular Theory of Gases and Liquids*, J. Wiley, London and New York (1964)

Hughel, T. J., *Liquids: Structure, Properties, Solid Interactions* (symposium), Elsevier, London and Amsterdam (1965)

Inman, D., and Lovering, D. G., *Ionic Liquids*. Plenum Press, London and New York (1981)

Jander, G., Spandau, H., and Addison, C. C., (Eds), *Chemistry in Non-aqueous Ionizing Solvents*, Vols. 1–4, J. Wiley, London and New York (1963)

Janz, G. J. and Tomkins, R. P. T., *Non-aqueous Electrolytes Handbook*, Vols 1, 2, Academic Press, New York and London (1972)

Jones, M. N., *Biological Interfaces*, Elsevier, London and Amsterdam (1975)

Kavanau, J. L., *Water and Solute–Water Interactions*, Holden-Day, San Francisco (1964)

King, E. J., *Acid–Base Equilibria*, Pergamon Press, Oxford (1965)

Kohler, F., Findenegg, G. H., Fischer, J., Posch, H., and Weissenbock, *The Liquid State*, Verlag Chemie, Berlin (1972)

Kratochvil, B., Gutman, V., and Smith, S. L., *Non Aqueous Chemistry (Topics in Current Chemistry*, Vol. 79), Springer, Berlin and New York (1972)

Kruus, P., *Liquids and Solutions*, Marcel Dekker, New York (1977)

Lagowski, J. J. (Ed.), *The Chemistry of Non Aqueous Solvents*, Vols. 1–5, Academic Press, London and New York (1966)

Latimer, W. M., *Oxidation Potentials*, 2nd edn, Prentice-Hall, Englewood Cliff, New Jersey (1958)

Linke, W. F., *Solubilities of Inorganic and Metal Organic Compounds* (data), Vols. 1 and 2, Van Nostrand, New York (1958)

Maitland, G. C., Rigby, M., Smith, E. B., and Wakeham, W. A., *Intermolecular Forces—their Origin and Determination*, Clarendon Press, Oxford (1981)

Malesinski, W., *Azeotropy and Other Theoretical Problems of Vapour–Liquid Equilibrium*, Interscience, London and New York (1965)

Mamankov, G. (Ed.), *Molten Salts, Characterization and Analysis*, Marcel Dekker, New York (1969)

Marcus, Y., *Introduction to Liquid State Chemistry*, J. Wiley, Chichester (1977)

McBain, M. E. L. and Hutchinson, E., *Solubilization and Related Phenomena*, Academic Press, New York and London (1955)

Morawetz, H., *High Polymers*, Vol. 21, *Macromolecules in Solution*, 2nd edn, Wiley-Interscience, Chichester (1975)

Munster, A., *Theory of the Liquid State*, North-Holland, Amsterdam (1965)

Nancollas, G. H., *Interactions in Electrolyte Solutions*, Elsevier, London and Amsterdam (1966)

Nyvlt, J., *Solid–Liquid Phase Equilibria*, Elsevier, London and Amsterdam (1977)

Ohtaki, H., and Yamatera, H., *Structure and Dynamics of Solutions*, Elsevier, Amsterdam (1992)

Parsons, R., *Handbook of Electrochemical Constants*, Butterworths, London (1959)

Pings, C. J., *Physics of Simple Liquids*, North-Holland, Amsterdam (1968)

Popvych, O., and Tomkins, R. P. T., *Nonaqueous Solution Chemistry*, J. Wiley, Chichester (1981)

Prigogine, I. (with Bellemans, A., and Mathot, V.), *The Molecular Theory of Solutions*, North-Holland, Amsterdam (1957)

Prigogine, I., and Rice, S. A. (Eds.), Non Simple Liquids, *Adv. Chem. Phys.*, **31**, J. Wiley, Chichester (1975)

298

Pryde, J. A., *The Liquid State*, Hutchinson, London (1966)

Reid, R. C., Prausnitz, J. M. and Sherwood, T. K., *The Properties of Liquids and Gases*, 3rd edn, McGraw-Hill, Maidenhead (1981)

Reisman, A., *Phase Equilibria*, Academic Press, London and New York (1970)

Rice, S. A., and Gray, P., *The Statistical Mechanics of Simple Liquids*, Interscience, London and New York (1965)

Rice, S. A. and Nagasawa, M., *Polyelectrolyte Solutions*, Academic Press, New York and London (1961)

Rigby, M., Smith, E. B., Wakeham, W. A., and Maitland, G. C., *The Forces between Molecules*, Clarendon Press, Oxford (1986)

Robbins, J., *Ions in Solution, an Introduction to Electrochemistry*, Clarendon Press, Oxford (1972)

Robinson, R. A., and Stokes, R. H., *Electrolyte Solutions*, Butterworths, London (1959)

Rochester, C. H., *Acidity Functions*, Academic Press, London and New York (1970)

Rowlinson, J. S., *Liquids and Liquid Mixtures*, 2nd edn, Butterworths, London (1969)

Rowlinson, J. S., and Swinton, F. L., *Liquids and Liquid Mixtures*, 3rd edn, Butterworths, London (1982)

Scatchard, G., *Equilibrium in Solutions and Surface and Colloid Chemistry*, Harvard University Press (1976)

Seidell, A., *Solubilities of Organic Compounds* (data), Vols. 1 and 2, Van Nostrand, New York (1941)

Shaw, D. J., *Introduction to Colloid and Surface Chemistry*, 3rd edn, Butterworths, London (1981)

Shinoda, K., *Principles of Solution and Solubility*, Marcel Dekker, New York (1978)

Sisler, H. H., *Chemistry in Non Aqueous Solvents*, Chapman and Hall, New York and London (1962)

Stokes, R. H. and Mills, R., *Viscosity of Electrolytes and Related Properties*, Pergamon Press, Oxford (1965)

Sundheim, B. R. (Ed.), *Fused Salts*, McGraw-Hill, Maidenhead (1964)

Tanford, C., *Physical Chemistry of Macromolecules*, J. Wiley, London and New York (1961)

Tanford, C., *The Hydrophobic Effect*, 2nd edn, J. Wiley, Chichester (1980)

Temperley, H. N. V., Rowlinson, J. S., and Rushbrooke, G. S., *Physics of Simple Liquids*, North-Holland, Amsterdam (1968)

Temperley, H. N. V., and Trevena, D. H., *Liquids and their Properties*, Ellis Horwood, Chichester (1978)

Timmermans, J., *The Physico-chemical Constants of Binary Systems in Concentrated Solutions*, Vols. 1–4, Interscience, London and New York (1959)

Ubbelohde, A. R., *The Molten State of Matter*, J. Wiley, Chichester (1978)

Vilcu, R. and Leca, M., *Polymer Thermodynamics by Gas Chromatography*, Elsevier (1990).

Waddington, T. C., *Non-aqueous Solvents*, Nelson, Walton-on-Thames (1969)

Waddington, T. C. (Ed.), *Non-aqueous Solvent Systems*, Academic Press, London and New York (1965)

Williamson, A. G., *An Introduction to Non-electrolyte Solutions*, Oliver and Boyd, Edinburgh (1967)

Zingaro, R. A., *Non-aqueous Solvents*, Heath, Lexington (1968)

Zundel, G., *Hydration and Intermolecular Interactions*, Academic Press, New York and London (1969)

Index